MOUNTAIN RESPONDER

MOUNTAIN RESPONDER

When Recreation and Misfortune Collide

Steve Achelis

Dog Ear Publishing
Indianapolis, Indiana

© 2009 by Steven B. Achelis
All rights reserved
Printed in the United States of America

No part of this publication may be reproduced, stored in a retrieval system, or transmitted, in any form or by any means, electronic, mechanical, photocopying, recording, or otherwise, without the written prior permission of the author. For information about permission to reproduce selections from this book, refer to the Contacts page on MountainResponder.com.

MountainResponder.com

Dog Ear Publishing
4010 West 86th Street Ste H
Indianapolis, IN 46268

ISBN: 978-160844-107-5

This book is printed on acid-free paper.

Printed in the United States of America

This book is dedicated to my daughters, Kimberly and Jennifer. Thank you for understanding that even when my commitment to the people in Salt Lake County took me away (so often at inconvenient times), you always came first.

Contents

PREFACE

My teammates and I saw and experienced things that few others ever will. This book is my attempt to share something of those experiences.

I selected these stories from a journal I kept during my time on the Salt Lake County Sheriff's Search and Rescue Team. Some of the entries were minutely detailed, right down to the color of a stocking cap or someone's choice of expletive; others gave nothing more than the date and the text from my pager. In these cases I had to glean details from news reports, other rescuers, and my own recollections. Although there are certain to be minor mistakes, the bigger picture is surprisingly accurate, in both spirit and substance.

Acknowledgments

Because this book is written from my perspective, I show up in it more than anyone else. That doesn't mean I did these rescues singlehandedly. Far from it. In fact, not one of them would have happened without the members of the Salt Lake County Sheriff's Search and Rescue Team—my teammates. Those skilled, self-sacrificing men and women, volunteers every one, deserve the credit for thousands of successful rescues. Although many of their names don't show up in this small sampling, I will always remember the experiences we shared.

To those of you who were rescuing others long before I joined the team, and to those of you who taught me the skills that helped me evolve from a backcountry recreationist to a backcountry rescuer, thank you; I treasure the lessons you taught me.

I'm honored to have served with all who gave their time so generously and asked for nothing in return. You're the real deal.

See you in the mountains.

Steve Achelis

ONE
CHILD IN RIVER…OR NOT

I'm helping my two daughters with their homework when my pager goes off. We share "looks." The timing, early evening, and the day, Sunday, are typical for a rescue callout.

I'm a newbie on the Salt Lake County Sheriff's Search and Rescue Team. Although I submitted my application in May of last year, interviewed to join the team in January, completed the background check in February, and passed the physical exam in March, it was May before I was actually sworn in by the sheriff. I still have to log several months of training and a nine-month probationary period before I'm a "full" member. It's mid-July, and I have responded to sixteen callouts.

I count to five to calm my nerves before reading the text on my pager, but it doesn't help—the adrenaline is already kicking in. What a newb! It's only an informational page about the Team's commitment to provide medical and traffic support at an upcoming marathon. About half our pages concern meetings or trainings. Maybe when I have a little more experience under my belt, the flood of adrenaline will slow to a drip, but I doubt that it will ever disappear—at least, I hope it doesn't.

An hour later, the pager goes off again—probably a correction to the earlier message, since that's usually what prompts a page that comes on the heels of an earlier one.

THIS IS A SEARCH AND RESCUE CALL OUT.
PLEASE RESPOND UP LITTLE COTTONWOOD
CANYON AT THE PIPE LINE FOR A
POSSIBLE CHILD FALLEN IN RIVER. EOM
21:04 07/22/01

I know this will probably be a fatality—we simply can't get there in the minute or two it takes to drown in a fast moving river. Jumping into my rescue clothes and boots, I grab a bottle of water and bolt for my truck. I radio into the sheriff's dispatch: "Salt Lake, Nine-twenty-five is ten-eight."

A woman's voice replies, "Nine-twenty-five at twenty-one-oh-nine."

Our communications are abbreviated to keep down the volume of radio traffic, although the shorthand does introduce some risk of miscommunication. In my call, the "925" is my identifier, it's actually called my car number—probably because cops are usually in cars, though I'm more likely to be on foot or skis. Car numbers in the 600s refer to sheriff's office deputies, and the ones in the 900s are search and rescue volunteers, like me. The 10-8 means I'm "in service." In their reply, they confirm that they received my transmission at 2109, which is to say, 9:09 p.m. Within the next few minutes, about a dozen rescuers sign in using similar jargon.

Climbing into my truck, I go through a mental checklist to see if I have everything I need for a swiftwater rescue: helmet, headlamp, life jacket, wet suit, gloves, booties, and throw bag. Check. I've made it a block down the street when my pager goes off again.

THIS IS A CORRECTION ON THE SEARCH
AND RESCUE CALL OUT. THIS IS NOT A
CHILD FALLEN IN RIVER. THIS IS A
FALLEN HIKER. REQUESTING TECHNICAL
CLIBMING GEER AND RESPOND UP LITTLE
AT THE PIPE LINE. EOM
21:11 07/22/01

I will come to learn that misspelling is to be expected in pages—the dispatchers are working to get the information out quickly, and they aren't going to spend precious seconds tidying up typos. The "EOM" is for "end of the message."

I will also soon learn that it's not uncommon to have a call change drastically, morphing into something virtually unrecognizable from the initial page. In this case, somebody heard cries for help near a river and called 911 to report a possible "child in the river." The caller likely pursued the screams and learned that it was a fallen hiker.

Although significant changes like this can be a little vexing, getting paged late is extremely frustrating. A year from now, I will be called to an SUV rollover in the mountains almost *two hours* after the accident. That's when the urban rescuers finally realize that they need assistance transporting the patient out of the canyon. That patient will die as we're hiking in.

As I drive up Little Cottonwood Canyon, radio traffic instructs us to take the Wasatch Resort entrance. After entering and driving a short distance, I come to a deputy who directs me toward a dirt road. I shift into four-wheel drive and jounce along through the dark woods for about a mile, till I see a fire department paramedics' vehicle, engine idling and doors open.

Stepping out of my truck, I see a sheriff's deputy whom I haven't met. She informs me that "923" has already headed up. I pull out my cheat sheet and see that 923 is Jon Blackburn. After a quick check of my backpack's contents, I step into my climbing harness and walk to a slope of loose rock, where Jon is scanning the darkness.

Jon is an experienced team member with a cheerful attitude. He has a habit of showing up at callouts with a three-day beard, which tends to annoy the brass. Jon asks a few questions to make sure I have the proper gear; then we turn on our helmet-mounted headlamps and head out.

Radio reports have been telling us there is "One ten-eighty-five Echo." The "10-85" means "patient condition," and "E," or "Echo" in the phonic alphabet used by cops and pilots, means it's an "obvious fatality." The radio also reported that two climbers are stuck on a rock wall.

Jon and I bushwhack up a steep hill through dense scrub oak and juniper, occasionally grabbing roots and branches to haul ourselves up the hill. Jon, whose now shredded legs are exposed by his shorts, is apologizing for the nasty route. I'm happy to be following, since I don't have a clue where we're going. We can hear car 606, a deputy, calling to us from above, and work our way toward his voice.

After about ten minutes, 606 yells down, "Have you reached the echo yet?" We haven't, and the thought unnerves me a little. I mention to Jon that I "haven't been around dead people very much." Not trying to be a wuss, but I want him to know that I'm not entirely sure how I will react. As a newbie, I figure I'd better be honest, because if I can't handle it, Jon will know soon enough anyway. Though it will seem a bit silly looking back later, my concern is that I won't see the body until it looms out of the darkness, inches from my face in the narrow tunnel vision of my headlamp, like something out of a B horror flick.

Ten paces later, I see the body. It is a young man, probably in his late-teens or early 20's. He's wearing only shorts—no shirt or shoes. His trendy "Joe Boxer" underwear strikes me as somehow ironic; he looks as though he might be a Joe. He is wrapped around a small tree, and I can't see his face. Initially, I take only a quick glance, to see how I'll react. My heart rate stays pretty even, so I look again. I feel sorry for him and his family, but those feelings aren't overwhelming. The stronger feeling is that I am glad to be alive. Of course, letting this thought into my consciousness brings a tinge of guilt. I'll learn a lot more about death, and the feelings surrounding it, in the coming years.

Maybe I've watched too much TV, but I ask Jon if we should cover him with a blanket. He replies, "No, ID will need to take pictures," and continues up the hill.

In another hundred feet, we come to 606 and a climber. The climber, a guide from Colorado, tells us he was teaching clients how to rock climb when he heard screaming from the wall above us. He tried to reach them but couldn't do it solo. He thinks he has a pretty good idea of how to get to the stranded people. He seems competent and asks if he can join us, but we have to turn him down—we can't add more unknowns to the mix. He understands.

In the United States, rock climbs are rated using the Yosemite Decimal System. A 5.4 grade is easy, and an experienced climber may not even rope up for it, whereas a 5.9 is my upper limit, leading roped, on a good day. This route looks like it's probably a 5.7 or 5.8, which is doable, but I certainly won't climb a 5.7 unroped, let alone at night on an unfamiliar route. Although we have ropes and harnesses, we don't have any climbing "pro"—wedge- or cam-shaped devices that fit into small cracks in the rock, to secure our rope should one of us fall. Jon and I agree neither of us is up for a nighttime 5.7 without pro.

The deputy shines a floodlight the size of a car headlight a few hundred feet above us. High on the vertical rock wall, we can see movement. He tells us it's one of the "victims," waving his shirt.

"Victim" is an interesting word, because it implies that an external force, somebody or something, hurt them, but everybody we help is a victim—typically of random misfortune or, more frequently, their own mistakes. A more politically correct, more innocuous term might be "subject," but then, that doesn't really capture the mood or the situation.

By their very nature, rescues tend to begin with some degree of disorder and confusion. But amid the often chaotic conditions, the responder has a very orderly set of tasks to accomplish: locating, accessing, stabilizing and transporting—the acronym LAST is the memory crutch and mantra quietly at work from the very outset. With the victims' location determined, we can cross the first task off our list; now it's time to access them.

Their pleas for help echo off the rock wall. One of them keeps yelling, "How long it will take?" The deputy cups his hands to his mouth and yells, "Hang on—Search and Rescue is on the way!" I think Jon knows as well as I that it's going to take us a long time to reach them. I sense the deputy's relief as he nonverbally hands off the responsibility to us, and I feel the weight of that responsibility settle onto our shoulders.

Jon and I begin climbing slightly different routes up the rock face, but we both soon reach near-vertical rock with poor handholds and have to retreat. Rejoining at the base of the wall, we agree to work our way right, to the west, along a narrow trail that may allow us to skirt the wall.

Jon is twenty-five yards ahead of me when the trail dissolves into a vertical wall. I try to work my way along the wall, but it bulges outward; below my feet, the ground drops off into blackness. Pressing against the sharp granite, my hands find reasonable purchase, but there's nothing for my feet. I back off and take a deep breath. I consider asking for a belay—having Jon anchor himself securely and pay out a rope to prevent me from falling too far—or telling Jon I don't want to go any farther, but I decide to take a better look. I won't let bravado dictate my actions, but I'm not ready to give up. I shine my headlamp where I'd like to put my feet, and see only blackness. Squatting down to get a better look, I push some branches out of the way and discover a sloping patch of dirt where I can place my foot. I again hold on to the wall, pressing my palms and the pads of my fingers against the crystalline granite, and slowly lower my foot into the pool of blackness until I feel the dirt patch, and edge my way around the bulge. I hope we don't have to do any more moves like that tonight!

Jon and I take turns leading through the brush. At times it's so thick we have to extend our arms out straight in front of us, spread the brush, and "swim" our way through the opening. There is some minor rock scrambling, where we need to use both hands and feet to climb, but nothing hairy.

Looking up, I see that our route is blocked by another vertical wall. There's a route to the left with easier handholds, but with pretty obvious slip-and-you-die consequences. Without voicing the thought, I know I won't go that direction without ropes.

To the right is a three-foot gap in the rock—a chimney going straight up about twenty feet. I ask Jon if I can try climbing it. He says okay, but I think he's a little concerned. This is our first time climbing together, and neither of us knows the other's skills.

I drop my pack and press my back against one side of the chimney, and my feet against the other. It's a good width, and I feel quite secure. I have Jon hand me the rope bag, which I push into a crack above me. I climb a few feet by inching my feet up, then my back, staying wedged between the two walls. Every few feet, I move the rope bag up a little higher.

As I work my way up the chimney, a helicopter hovers a few hundred feet off to my side. They are shining an enormous spotlight—I think they call it the "night sun"—on me. I'm sure they are trying to help, but the deafening *whop-whop-whop* of the main rotor may cancel out the benefit of enhanced visibility.

I feel as if I'm in a strange dream. It's almost midnight, and I'm soloing up a rock chimney, lit up by a helicopter, in an effort to rescue two stranded climbers. There's a dead guy in the trees below me. It seems I should be heart-thumping excited, but I feel strangely calm.

After several minutes of squirming, I'm through the chimney and scrambling up seventy feet of easy rock. I set an anchor around a foot-thick pine tree. Knowing that Jon is unsure of my skills, I make the anchor textbook perfect: two independent anchors, two locking carabiners, on a bombproof tree. It's complete overkill, but I expect he'll check it when he gets up the pitch.

I tie into the anchor and throw down one end of the rope to Jon. We exchange the standard commands over the radio: "On belay?" "Belay's on!" "Climbing." "Climb on!"

All of a sudden, other traffic swamps the radio and drowns out any chance of our communicating. Didn't they hear us go on belay? In hindsight, I should have called for radio silence or switched to a different channel. I have much to learn.

Jon is climbing slowly, and I can't tell why. Is he unsure of the route? Unsure of my anchoring and belaying skills? Is the chimney too narrow for him? After ten minutes, he's finally out. It turns out that the reason for his slow progress was that he was carrying both our backpacks. Blush.

Over the radio, we learn that a team of four rescuers is working their way toward us. Jon considers leaving our rope so they can come up by tying a special friction knot, called a Prusik, onto the fixed line; the Prusik slides up the fixed rope as the rescuer climbs, but it grabs the rope if he falls. We quickly agree that belaying is faster and safer—and we don't want to leave our rope behind.

I belay the first member in the next team. We then trade out ropes so he can belay the rest of his team while Jon and I pack up and head higher.

Ten minutes later the helicopter contacts us by radio. "Climbers on the mountain, this is Life Flight."

"Go ahead, Life Flight."

"The stranded party is below you and to the east."

That's good news. We're getting closer. Jon explores the terrain and finds a narrow couloir to our east. Unfortunately, neither of us have any of the bright-orange surveyor's tape we typically use to mark our route for the next team. Instead, I inflate a surgical glove until it looks like a huge, white cow udder, and tie it to a bush at the turnoff. I use my radio to inform the second team that our turnoff is marked with a glove. I'm sure they have no idea what I'm talking about. I'm also sure they will know exactly what I meant when they see the inflated glove.

After climbing the couloir and traversing a plateau, we learn from the helicopter pilot that we are directly above the victims. I ask how far below us they are. "About seventy-five feet," is the reply. I then ask how far the victims are above the deck. They reply, "A long, long way." Jon and I agree that seventy-five feet plus "a long, long way" exceeds the length of our two-hundred-foot rope. It might even be longer than two ropes.

Team 2 has arrived, and rescuer Alan Erdahl walks up to the sloping edge and peers over, looking for the victims. He's comfortable standing on the edge of the three-hundred-foot wall, but it is unsettling to watch. Jon speaks up with puzzlement in his voice, saying, "Erdy, what are you doing?" Erdahl gives him that well earned "I've been on this team since before you were born" look, and it's true—he has been doing rescues for thirty-six years, but I'm happy to see him stop and set up a safety line.

The standard method to "pick off" a climber who is stranded mid face is to lower a rescuer to the climber and then lower both rescuer and climber to the ground. But we need to pick off two people, and our two-hundred-foot ropes won't reach the ground. Instead, we'll lower a rescuer to the victims and then raise them to the top of the wall.

With the helicopter hovering above us and flooding us with noonday sun, the six rescuers set up three rope systems. One of the ropes will be a lowering system, which I will use to lower Jon over the edge to the stranded climbers. The second rope will run through several pulleys, which will give us enough mechanical advantage to raise the climbers. The third rope will be a belay—if something goes wrong with the other systems, the belay will catch them.

Jon ties into the rope and walks backward until his feet are on the edge and his back is hanging over the abyss. He looks up, and all I see is his bright headlamp and his silhouette. "Down slow" he says, and he slowly disappears from sight. It's a little like having a big fish on the end of the line—all I see is the taut line running into a pool of darkness, though I know Jon is still there by the changing tension on the rope.

After several minutes of watching the rope slide over the edge, we hear Jon's voice on all five radios: "Stop!" He tells us he's at the same height as the victims, but they're about fifteen feet to his left.

The climbers are standing on a precariously tiny shelf about half the size of a doormat, in a body-size crack in the rock wall. They're holding on to a single gnarled branch growing out of the rock. It's doubtful that the skinny branch will support either of them, but it gives them just enough stability to keep them from falling out of the crack, Jon can see the fear in their eyes. They just watched their friend, brother to one of them, fall past them and into the void below.

Though I can't hear the conversation taking place a hundred feet below me, I know what Jon is saying as he approaches the young men: "Don't touch me; don't grab me; don't *move.*" Strong words, intended to let them know that he is in charge and to keep them from reaching for him. It wouldn't be the first fatality resulting from an overeager victim's ill-timed lunge toward safety.

To reach them, Jon uses his feet to "walk" across the wall to his right. He then "runs" back to his left so he can pendulum his way over to their location. Reaching out, he grabs the rock and holds himself in position. He will later tell me that he wished he had some protection so he could secure himself to the crack. As it is, he will have to hold on with one hand and work with the other.

The victims have now been located and accessed. Jon's new task is to get them stabilized so they can't fall off the cliff.

He pulls two runners, loops of webbing that he has draped over his neck and shoulder, and hands them to the men. Working with one hand and giving instructions, he has them each put one arm through a runner, reach behind their backs with the other arm, and pull that arm through, so that the runners are behind their backs and looped over the front of each shoulder, to create chest harnesses. He then has them clip the ends of the runners on the front of their chests with locking carabiners. Then, using Prusiks and additional runners, he fixes the two men to one of the ropes. He then takes another piece of webbing and creates an improvised "diaper" harness for the victim closest to him.

As the minutes pass, our anxiety is growing on the top of the cliff. Unaware of Jon's one-handed cats-cradle project, we don't understand why it's taking so long. Erdahl, who is now secured by a rope and leaning over the top of the cliff, uses his radio to ask Jon his status. Jon says he's okay, but I can hear his frustration with our repeated radio calls, which are interrupting his already slow progress.

After connecting one of the stranded climbers to the raising and belay ropes, we finally hear the words we've been longing to hear: "Up slow!"

Two rescuers pull on the rope that is threaded through the pulleys. Theoretically, the pulleys should make the load feel one-third as heavy, but because of the friction, rescuers still have to pull more than half the victim's actual weight. With each strenuous pull, they yard in about two feet of rope, and the victim rises eight inches.

The young man, who is now dangling two hundred feet up above the valley floor, must be puzzled by the slow progress. Rise eight inches, only to pause and bounce on the rope. Then another eight inches and another bounce. After raising him fifteen feet, we have to stop and reset the pulleys. The eight-inch increments restart until another fifteen feet have passed. We do this repeatedly until at last we get him up to the edge.

As he comes over the top of the cliff, reality begins to sink in. He looks tired, wobbly, and emotionally drained. I meet him at the edge and help walk him up the slope. As I do, I ask him if he's okay. He looks me in the eye and says, "Fuck no—my brother just died!" His pain hits me like a rock between the eyes.

His next words are, "Does anyone have a smoke?" I doubt that he realizes the irony of asking mountain rescuers for a cigarette. Then again, I might want one, too, if I were in his situation.

It's difficult to imagine what these young men were thinking as they began their untethered assault on this rock wall. Then again, with the testosterone-fueled immortality of youth, they probably weren't doing much thinking.

With the first victim safe, Jon can now move into the crack with the remaining victim. He pulls down the ends of the ropes from above, attaches them to the makeshift harness on the second man, and again gives the "Up slow!" command. I'm sure every one of us wishes we could bring him up faster, but all we have is the gear we could carry in: a few ropes, carabiners, slings, and pulleys.

The second man rises like the first, eight inches at a time. When he's safely up, we send down the ropes for Jon and repeat the strenuous process.

As Jon crests the top, he definitely has his game face on. He's been on point, hanging it out for almost two hours. As he moves up from the edge and receives a few congratulations, I can see a wisp of a well-earned smile sneak over his face.

The two young men sit with their backs against a large boulder, knees drawn up and heads hanging low, as we dismantle our mass of gear. One of the victims asks if anyone has a phone. Rescuer Chris Patch, in an effort to be helpful, says yes and goes to his backpack. Bob Mead, a newbie like me, goes over to Chris and murmurs, "The guy's brother just died and we're still on the mountain. Do you think it's appropriate to have him make a phone call?" Chris tells the young men that his phone isn't getting a signal—an appropriate white lie.

The young men are now off the rock face, but we still have to get them safely off the mountain. As we descend, we spot a couloir that looks like a shortcut. Another rescuer and I climb down some two hundred feet until we are halted by a cliff that, at least in the dark, appears bottomless. Climbing back up and right, we find the tree we used as an anchor above the chimney. Rescuers rappel down the chimney, and the victims are lowered.

Below the chimney, we come to the bulge that stumped me on the way in. Jon and another rescuer have rigged a twenty-foot hand line across this span. After the traverse, we head down a game trail. Halfway down, my headlamp finally winks out—it's been on for five or six hours. If I leave it turned off for a few minutes, it will work for fifteen seconds before going black again. Using this technique and staying between two teammates, I finish the descent.

We started at nine thirty p.m. By the time we finish a quick debriefing and I drive home, it's four thirty a.m. Ironically, my alarm clock went off at three thirty so Bob Mead and I could get an early start on a morning hike. We'll pass.

Two
Busy Weekend

July 4, 2003, fell on a Friday. I had been on the team for two years, and I was sure we would be called out. The summer months of June, July, and August are definitely our busy season, when we get more than half our calls, and numbers nerd that I am, I had determined that based on callouts over the past seven years, there was a 75 percent probability of a callout on the Fourth of July. It didn't happen.

Sitting on my porch overlooking the Salt Lake Valley, I watched the fireworks and went to bed.

Stranded in Butterfield Canyon

I wake to the incessant beeping of my pager. Looking at the clock next to my bed, I see that it's two thirty a.m. Bleary-eyed, I turn on the lights and squint to read the tiny green screen:

```
S/R callout cars to meet at the mouth
of Butterfield cyn. Cars need to
bring High angle gear and mountain
rescue, this is on cd 1
02:23 07/05/03
```

Calls to Butterfield Canyon are frustrating, because it's on the other side of the valley. It will take me about thirty minutes to drive to Herriman, and another fifteen to drive up the Canyon. Most of these calls are handled by County Fire, which has a station in Herriman, long before we arrive.

Every call that I have responded to in Butterfield Canyon has been for cars that have driven off the winding dirt road. My last call there involved a truck that rolled off the road but didn't make it down to the next switchback. We rappelled down from above; I remember looking at the dead teen surrounded by shiny metal beer cans and CDs. This may well be a similar call.

I pull on my rescue clothes and walk to my truck, turn on my radio, and tell the dispatcher I am en route: "Salt Lake, Nine-oh-five [my number at the time] is ten-eight." Her reply is terse: "Nine-oh-five at two twenty-five."

During my drive, the radio gives periodic updates. A truck has driven off the road; both occupants were ejected from the vehicle—never a good sign. One is "ten-eighty-five alpha," which means unhurt, and the other is "delta"—critically injured and possibly dying. It doesn't take much to slip from delta to echo. Additional updates inform us that Fire is on scene. Patient vitals, blood pressure, and pulse are given for the delta patient. They don't sound good.

A repeat page is sent twenty-eight minutes later. This means that not enough rescuers have responded and that incident command is concerned that they will need more people.

The on-scene Fire personnel report that the patients are on a "seventy-five-degree" hill. That's steep and will definitely require vertical rope-rescue gear.

As I near the canyon, I hear that Life Flight, one of the valley's two air ambulance services, has landed and Fire is transporting the patients. *What?* How did they transport the patients on a seventy-five-degree hill? Oh, well, in the heat of the battle it's easy to overestimate the terrain.

Several miles up the canyon, I see the helicopter parked in the middle of the road, and a few sheriff's office vehicles with lights on. I pull over to the side next to a fellow rescuer's vehicle. We know this will be over soon. At three thirty a.m., we hear the message over the radio, "You can cancel Search and Rescue." The message is bittersweet. It's a little frustrating to respond to a scene where our services are not needed, but not as frustrating as being called late. In this case, the patients' best interests were met: they were quickly packaged and transported to the air ambulance. Although I don't like waking up at two thirty in the morning to drive across the valley, I understand why we were called and then not used.

The final page comes at 3:44 a.m.:

```
CANCELLATION FOR SAR CARS ON THE
ROLLOVER IN BUTTERFIELD CANYON. EOM
03:44 07/05/03
```

I drive home and climb into bed. Of course, getting into bed is one thing, and going to sleep something altogether different. Even a call as boring as this leaks a little adrenaline into my system. I try to relax by picturing myself making perfect telemark ski turns in fresh powder.

Cliffed Out on Mount Olympus

Fifteen minutes later, thirty pagers light up simultaneously.

```
S/R callout for overdue hikers. Cars
to meet at the Mount Olympus
Trailhead, flares are needed, this
will be on Cd1. EOM
04:45 07/05/03
```

Another jargony message. The term "flares" does not refer to the bright red roadside pyrotechnics you might expect, but to an acronym for forward-looking infrared sensors (FLIRS)—a night vision system that is used on the Utah Department of Public Safety (DPS) helicopter. "Cd1" refers to the radio frequency we will be using: code one. Mount

Olympus is Search and Rescue's nemesis—an immense mountain rising 4,400 vertical feet from the Salt Lake Valley.

Five minutes later, at 4:50 a.m., a repeat page is sent. This is a reminder that, tired though we may be, we are needed. There is no shortage of business when it comes to accidents.

I'm the first rescuer to arrive; I get a quick briefing from a deputy. We have two missing people: an eighteen-year-old male and a fifteen-year-old female. They hiked up Mount Olympus to watch the Fourth of July fireworks and told their parents they would be home by two a.m. The girl's mother left the trailhead to search for her daughter a half hour ago. The deputy called the mother on her cell phone and asked her to come back to the trailhead, but she refused. That's understandable—I would probably want to look for my daughter, too. Still, it complicates the search, because we now have three people on the mountain.

I quickly load my backpack. I'm assigned to the "hasty team," which means we will be moving fast and carrying minimal gear. I take my radio, helmet, several liters of water, and a handful of energy bars. I attach a bright-green glow stick to my backpack.

Another rescuer, Dan Smith, a.k.a. car 922, arrives as I am packing. Dan has been on the team for six years; when he isn't rescuing people, he's running his local tire store. He, too, assembles his gear. In the dark, with packs and headlamps on, we take the first steps up the steep Mount Olympus trail. It will be a long night. I press the stopwatch button on my wristwatch. Within minutes, sweat is coming through my clothes.

The radio traffic reminds us that we are not alone as additional searchers arrive, are equipped, and are deployed.

At 5:33 another page is sent, requesting that more "cars" respond. It has been forty-five minutes since the first page. We have three two-person teams headed up the Mount Olympus trail, and one team headed in the Heughs Canyon trail. Dawn is breaking as we turn off our headlamps.

Hiking as a rescuer is much different from hiking for recreation. First, you don't get to pick when you start—the hike begins with the sound of the pager. Second, you rarely stop to rest along the way. If you're thirsty, you stop only long enough to swallow some water. There is also a subtle level of competition to hike fast. Some of this is simply human bravado—nobody likes to be passed—but most of it comes from wanting to help the "victim" as soon as possible.

The sheriff's "command post" arrives at the Mount Olympus trailhead. This is a motorhome equipped with a TV, computers, radios, rescue gear, and refrigerators full of water and sports drinks.

The Search and Rescue member orchestrating this mission is car 920—Darren Westerfield, one of three elected squad leaders, and a personal trainer by day. His job is to make the decisions on which team members and equipment to deploy, and where.

Darren is in cell phone contact with the mother, who has managed to get herself lost on the mountain. She no longer thinks she is on the main trail. She tries the next-to-impossible task of giving her location by describing her surroundings.

One of the challenges Darren faces on this early morning rescue is whether to focus his limited resources on the mother or the overdue hikers. Although the hikers probably need our services more urgently, the mother will probably be easier to find, which will then free him to focus all his resources on the overdue hikers.

The incident commander at the command post has arranged for a helicopter, which made one pass with its night vision FLIRS system, but now it's too light to use the "flares" effectively. The spotter in the chopper sights the mother, more than an hour up the trail.

As we are hiking up, we call out the names of the mother and the daughter and pause to listen for a response. After about an hour of hiking, we hear the mother's reply, and we see her a few minutes later.

We wait with the mother until the second team arrives to escort her down. She's fit and able to walk down on her own, but we don't want her getting lost again.

Dan and I continue up the section of the trail affectionately known as Blister Hill. It is six thirty a.m., and we're an hour and a half into the rescue.

The helicopter lands to add rescuer Craig Miller to the airborne search. Within a half hour, Craig announces that he has spotted the lost hikers. They're in Heughs Canyon, a half mile south of our position, stranded by near-vertical terrain above and below them. Craig says they look healthy and mobile. A 150-foot cliff band has halted their downward progress.

The location doesn't surprise us—it's a common mistake to slip into the Heughs Canyon drainage while descending Mount Olympus. Unfortunately, the upper reaches of Heughs Canyon are rugged, rarely traveled, and very steep.

A brief radio discussion ensues about whether to use the helicopter to transport additional rescuers, or just to summon Life Flight's chopper and use its hoist to extract the hikers. The pilot reports that there are no safe LZs—landing zones. It is quickly agreed that our apparently healthy victims don't justify the added risk of a helicopter hoist operation. In this mission we will limit helicopter use to an observation platform—the rest gets done with boot leather.

To conserve fuel, the helicopter must land on Wasatch Boulevard several times, so sheriff's deputies close the road from both ends. No doubt, people driving on the boulevard find the repeated road closures annoying, but maybe they pause and imagine if it were their son or daughter stranded high on a cliff.

Normally, a half-mile hike is nothing—a casual few minutes for any fit person—but this is in upper Heughs Canyon, with some rugged terrain between the stranded party and us. We have several options: (1) head south from our current position and traverse a series of steep scree slopes and rock walls; (2) continue to the saddle near the top of Mount Olympus and then work our way down into Heughs Canyon using the route our hikers took; or (3) send more rescuers into Heughs Canyon, approaching up from the base. Darren chooses a combination of all three.

Dan and I, designated Team 1, begin the over-hill-over-dale approach heading south from our current position. Rob Hunter and Shawn Rowland, Team 2, will continue up the trail to the saddle, and Alan Erdahl and Andy Peterson, Team 3, will split up, with Alan escorting the mother down and Andy joining Team 2. Rescuers Alan Bergstrom and Gary Nelson, Team 4, are coming up Heughs Canyon. Additional teams are formed and sent into Heughs Canyon.

When Dan and I reach the first ridge between the Mount Olympus trail and Heughs Canyon, we begin to see our challenge. The terrain is very steep and rocky. Using his GPS, Rob calculates a compass bearing of 101 degrees between Team 2 and our stranded hikers. Dan and I locate Team 2's position on a map, draw a line at 101 degrees, and mark an "X" where the line crosses the 7,100-foot contour—the elevation estimated by the pilot. The line crosses numerous steep ravines; this will be interesting.

Dan and I spend the next two hours traversing steep gullies filled with loose rocks, interspersed with moderate rock climbs. Sgt. Lane Larkin, the incident commander with the sheriff's office, and Darren, the operations commander on this rescue, board the helicopter to get a bird's-eye view of the operation.

Sergeant Larkin has been the liaison and the incident commander of the Search and Rescue Team for more than a decade. Lane is something of a father figure, pushing us to excellence yet also looking out for our well-being and safety.

When I hear Lane's voice on the radio, I know we are in good hands.

From his vantage point in the helicopter, Lane helps us with route finding by calling in landmarks: "Traverse to the dead tree in the center of the next ravine, climb the rock to the first notch near the ridgeline, and then stay near the line of brush as you traverse into the next ravine."

The helicopter spotters also guide Alan Bergstrom, who is heading up Heughs Canyon, by pointing out landmarks. He soon reaches a location adjacent to the hikers, where he establishes voice contact. They are thirsty but unhurt. Alan reports that to go any farther he'll need climbing gear and a belayer. The helicopter returns to the command post, gets climbing gear, and, from a low hover, drops bags containing several climbing harnesses and two two-hundred-foot rescue ropes to Alan.

The teams on the mountain are well positioned. Alan stays in voice contact with the victims, though he can't proceed without a climbing partner. Dan and I are still thirty minutes to an hour away from the victims. The three members of Team 2 are at the saddle near the top of Mount Olympus. Team 2 will serve as a backup team in case anyone is injured during the rescue. Two additional teams are advancing from the bottom of Heughs Canyon with more gear and water.

Lane informs me that we are above the stranded party. Unfortunately, we can't reach them without ropes, and Alan, who has the ropes, can't reach us without a climbing partner. I down-climb as far as I can toward Alan, then tie several pieces of climber's webbing together and lower the forty-foot strand to Alan. He ties the climbing gear and a rope end to the webbing, and I haul it up the rock face. Once we have the gear, I use the rope to belay Alan from above as he climbs the face, while Dan sets up an anchor so we can rappel down to the stranded kids.

When we reach them we find that they're in good condition. Dan talks to them lightheartedly and tries to inject a little humor to keep them calm. Of course, humor is a bit hard to come by, because we have haven't slept since our Butterfield Canyon call at 2:23 a.m.

After setting up a bombproof anchor on a nearby rock, Dan rappels down the face and connects the second of the two ropes dropped to Alan from the chopper. Alan and I pull up this second rope. We will need both two-hundred-footers to safely lower our hikers and retrieve our ropes.

Both our teenage hikers have rappelled before, so we cover the basic lowering instructions quickly. We'll lower them one at a time. They are to lean back on the rope and walk their feet down the rock wall as if they were rappelling. I feed the rope through a brake bar rack that controls the speed of their descent. Both kids are physically coordinated and follow instructions well.

When the five of us have descended the wall, we create an anchor for our next rappel. We then repeat the lowering/rappelling process two more times until the five of us reach the fresh teams that have been sent up Heughs. We send our two "guests" out with these teams and begin packing up our gear.

From the base of Heughs Canyon, we are shuttled back to the Mount Olympus trailhead in the deputies' SUVs. I can see Sergeant Larkin, surrounded by a half-dozen TV cameras; their lenses are within a few feet of his face as he fields the reporters' questions. I glance at my watch. It has been eight hours and forty-five minutes since I left the trailhead. I'm ready for a snooze.

I get home, taxi one of my daughters to her friend's house, take a shower, and climb into bed. It's four in the afternoon, and sleep won't come—even the repetitive left turns of a NASCAR race on TV can't lull me to sleep. I get up, run some errands, and hit the sack at nine p.m.

Cramped on Lone Peak

An hour and a half later, my pager goes off:

```
S/R callout for assist for Draper PD.
2 dehydrated hikers poss by Jacobs
Ladder. Draper Sgt. on scene, no
further info at this time. This is on
our Cd1 freq. Eom.
22:27 07/05/03
```

I sit on the edge of my bed looking at my pager. *Crap.* Jacob's Ladder is a section of trail on Lone Peak, the 11,253-foot monster that towers above the south valley. Lone Peak rescues typically involve the use of ATVs and long hikes—sometimes twelve hours or more. I know there are thirty other rescuers sitting on the edge of their beds looking with blurry eyes at their pagers. I hope some of them managed to get in a catnap today.

Driving to Corner Canyon, the radio traffic is almost overwhelming. There are several agencies working this rescue. Sandy City Fire has already headed in. Draper City's new police department, which has existed for only five days, is on scene. The Salt Lake County Sheriff's Office, which has legal responsibility for all search and rescue missions in the county, is in the awkward position of listening to the agencies that have already begun the search. Life Flight has been called, as has the DPS helicopter. The radio traffic is confusing and contradictory—nothing unusual or unexpected in a multiagency operation—though I expect things will settle down when a single incident commander emerges. Sergeant Larkin, who has probably had less sleep than I, begins coordinating the interagency communication while Gary Banks takes over the rescue's operational command.

Gary, a wiry guy with a white goatee, is the commander of the sheriff's Search and Rescue Team. He has been on the team for 25 years, has participated in more than 1,200 rescues, and has been elected commander by the team for each of the past 10 years. As commander, Gary is responsible for all the team's activities, and he prefers to do his leading from the field rather than the command post. He had just returned from vacation when this call came in.

Gary has been listening to the almost nonstop radio traffic between Sandy Fire's two teams and the Life Flight helicopter. He's on the phone with the victim, who reports that he spent last night camping in the Lone Peak cirque, did a rock climb to the summit, and was hiking out when he was overcome by fatigue and terribly painful leg cramps. He's about thirty minutes up the trail in his sleeping bag, and he's cold. Gary talks him into getting his flashlight, which takes significant effort on our patient's part, and shining it up at the Life Flight helicopter. Life Flight sees the light and relays the victim's GPS coordinates to the rescue teams.

Although it takes only half an hour to hike to the patient, carrying him out on the litter will take almost two hours. The trail is narrow, and the litter is heavy and crowded with the patient, oxygen bottle, vitals monitor, and IV bags. We stop only to switch out litter carriers, occasionally set up a rope to help us navigate a steep section of trail, or crush a stinging red ant that lands on our patient.

On the way out, we meet a team member who has brought in the "wheel"—a big, soft tire from an ATV (All Terrain Vehicle) which we will attach to the litter's aluminum frame. Although rescuers must still balance the litter, the wheel supports most of the load as it rolls along the trail. As we descend the steeper sections, three team members manage a belay line attached to the litter, so it can't get away from us.

At the bottom of the trail, the patient is quickly moved to an ambulance. We then go through the process of reorganizing, restocking, and repacking our equipment. It's four in the morning, and I've had a little more than an hour of sleep since two a.m. yesterday—and much of that time has been spent climbing or hiking.

When I get home, the sun is rising, but this time I have no trouble falling asleep.

Rockfall on the Pfeifferhorn

I wake up after five hours of black, dreamless sleep. The past thirty hours have been physically demanding. Although I'm still tired, I know that a gentle workout will reduce the lactic acid in my muscles and ease the inevitable muscle soreness. I go to the basement, lift a few weights, and spend twenty minutes on the Stairmaster. When I return upstairs, I remember that I left my pager in the bedroom. At a glance, I see that it went off while I was exercising:

```
SEARCH AND RESCUE CALL OUT ASSIST
HIKER LIL CTNWD CYN. RED PINE, CONT
610 ON CODE1. CS 78334 EOM
11:43 07/06/03
```

I throw on rescue clothes and drive up Little Cottonwood. During the drive, I hear that Life Flight has landed at the Red Pine trailhead. Cars driving up or down the snaking two-lane canyon road have been stopped, creating a large traffic jam that is limiting the rescuers' ability to get to the scene. Deputies consider having the inbound rescuers drive up the now-closed downhill lane, but they decide it's safer to wait for the road to reopen.

Early reports by cell phone inform us that our victim was near the summit of the 11,300-foot Pfeifferhorn when she was hit by a fourteen-inch rock, dislodged by a hiking party above her. Imagining that fifty–something-pound rock tumbling down the hill, I anticipate serious damage, inside and out. Fractured ribs? Internal injuries? The victim, an emergency room physician, gives a preliminary self-diagnoses of her injuries: sure enough, fractured ribs, and possibly a ruptured kidney.

Due to the elevation and wind conditions, Life Flight can't lower rescuers directly to the patient. Instead, Gary Banks and a Life Flight paramedic are lowered onto the ridge below the summit; from there, they reach the patient after a short hike.

Gary calls for six more rescuers and specifies the rescue gear we'll need. Darren Hunsaker, currently the team's vice commander, is running this operation from the trailhead. Unlike most of us, Darren joined the team with minimal climbing or mountaineering experience, but he threw all his considerable energy into mountain rescue and is now a widely respected veteran. He organizes the rescuers into three teams of two. We are told to gather our gear fast and prepare to be flown in. Without the helicopter, it would take five hours to hike to the summit with our rescue gear, and at least twice that time to carry the patient out. Indications are that she may not survive a fifteen-to-twenty-hour rescue.

As we are assembling equipment, the Life Flight paramedic completes his medical assessment and classifies her as a priority patient. Gary and the paramedic immobilize the patient in a full-body vacuum splint, or "beanbag," and call for the helicopter. Paramedic and patient are hoisted up to the side of the hovering helicopter at 11,000 feet, and flown to the trailhead, where they are lowered to the road. There team members assist in transferring the patient from the helicopter's hoist to a gurney as the helicopter lands a hundred feet away. We then secure the gurney inside the helicopter, which flies her to the hospital. She is angry and combative, swatting at anyone who tries to help her. I suspect this is due to her injuries, but it may just be that she's used to being the one giving orders and doesn't quite know how to be a patient.

While the helicopter is transferring her to the hospital and then refuels, Gary hikes down the mountain to find a suitable LZ where he can be picked up.

Geckoed in Mill Creek

In the last thirty hours, we've had four rescues: from a car rollover on the western edge of the Salt Lake Valley, two cliffed out hikers and a woman injured by a tumbling rock on the east, and an exhausted climber on Lone Peak in the South. Before the day is over, we have one more canyon to visit: Mill Creek.

My pager sounds at 6:30 p.m.:

```
This is a callout for S&R on a fallen
hiker at Mill Creek Canyon. Pipeline
trail. Respond on Code One. Eom
18:33 07/06/03
```

Glancing at my pager, I realized that this may be a difficult victim to locate, because the Pipeline trail stretches more than three miles along the north side of Mill Creek canyon. The radio gives the preliminary details. There are two victims. One is a climber who has already fallen. Her injuries sound serious. The other victim is now hanging on to the rock wall and calling for help; she doesn't think she can hold on much longer.

Few of our rescues are so urgent that a few minutes will make the difference between life and death. This one may. Our immediate challenge is to locate the victims along the three miles of Pipeline Trail. Radio traffic reports that Fire paramedics are hiking in with the witness.

Darren Hunsaker, Dan Smith, and I arrive at the trailhead within minutes of each other. We know the urgency of this call. Dan quickly stuffs a rescue rope and an assortment of climbing hardware into his backpack. As soon as he is loaded, he starts up the trail at a fast pace. I put on my climbing harness, grab a full-body vacuum splint, and follow him. Darren, who is running this rescue, wraps up his preliminary operational command duties, grabs more climbing gear, and heads in after us.

When we arrive at the Pipeline Trail junction a few minutes later, the three of us know our immediate dilemma. The trail heads both east and west, and the Church Fork trail continues north. Which of the three directions should we go? I try to contact the Fire dispatcher on the radio and hear her faint reply, "…Sheriff…," then a barrage of static. The steep canyon walls are blocking the signal. Dan and I decide to head up the steeper Church Fork trail, since it's a likelier spot for climbers and because we know we can come back downhill fairly quickly. Darren stays for a few minutes to handle radio communications.

Within five minutes, I hear Dan yell that he has the victim in sight. When I arrive, I see that Fire is already working on the first patient, who is at the base of the rock wall. Recognizing several of the paramedics as being from Station 12, I know she's in good hands.

Looking up, I can see a young woman forty feet up a vertical wall, clinging like a spider—no rope, just hand and foot friction. She's calling for help and tells us, "I can't hold on any longer!" Dan asks the bystanders to be quiet and then tells the young woman, "Everything's going to be okay. Your job is to relax and focus."

A bystander points to a small trail that leads above our victim and mentions that there are "climber chains"—preset anchors—near the top. This simple bit of information saves us several precious minutes.

It takes only a minute to reach the top of the sixty-foot wall. We can see the climbing chains just over the edge. I tie a large loop into the end of the rope for our victim and clip a smaller loop into my harness. Dan throws a section of the rope behind his back for a hip belay, braces his feet against a big rock, and lowers me over the edge.

When I reach the chains, I clip my personal tie-in to one of the bolts. I then tie the rope to a carabiner using a Münter hitch, which takes most of my weight, letting Dan control my descent easily. Darren arrives and throws me the other end of the rope, which I clip into. This will be a "tag line" that he can use to "pendulum" me over to the victim if needed. Now I can free myself from the bolt anchors and be lowered to the victim.

As Dan lowers me, long-term Search and Rescue member Rob Webster is below her. In a calm voice, he tells her, "A rescuer is coming to help you. It is very important that you do not touch or grab him. Okay?" He uses his radio to relay commands to the team that is lowering me.

As soon as I am alongside the young woman, I take the long loop that I tied earlier, and pass it between her legs and around her hips, fastening it with a locking carabiner, as if it were a big safety pin on the front of a diaper. This makeshift harness will not be comfortable, and it certainly isn't something I've used when practicing a "pickoff," but it's a quick improvisation that will keep her from falling.

Once she is stabilized, I wrap a Prusik onto the rope above me. I then take a small runner and connect it between the Prusik and her improvised harness. By moving the Prusik up or down, I can adjust her height so that we are hip to hip, rather have her dangle below my harness.

With her now suspended next to me, Rob uses his radio to give the "down slow" command to the team above me.

As soon as we reach the ground, and while rescuers are freeing her from the improvised harness, she is overcome with emotion and begins sobbing. She watched her sister fall almost thirty feet, and she has been stranded on the rock face for more than an hour.

At first, this looked like a case of inexperienced climbers trying, unroped, to get up a crag that "looked doable." Committing a common rookie error, at some point they made an upward move that they could not then safely retrace. Thus, they found themselves committed to "topping out"—the only way to go was up, into yet thinner and scarier terrain. But as stiff as the climbing is, I suspect that they were experienced boulderers who simply got in over their heads (literally), with much the same result.

Before this weekend, I was certain we would have a call-out on July 4 and was surprised when we didn't. I was even more surprised to have nine victims in five callouts over the next forty hours.

Three
Waterfalling

The call on June 30, 2004 reports that three children in Bell's Canyon are suffering from severe dehydration. We have two air ambulances inbound when rescuers reach the hikers near a dirt road. The party is tired and thirsty but they are not experiencing a medical emergency.

While this is going on, we get a call on a fallen climber in a waterfall just a few miles from our current staging area. Although Utah is an alpine desert and we're in a multiyear draught, this call will turn out to be one of seven waterfall rescues during the summer of 2004.

We dispatch most of the team to the second call and divert one of the inbound medical helicopters.

Gary Banks, Dan Smith and I are the hasty team. It takes us less than twenty minutes to reach the base of the waterfall. I can feel the wind from the helicopter hovering a hundred feet above us. The sound makes talking difficult.

A young woman is standing near the base of the falls. Over the beating roar of the helicopter, Gary asks her, "Is he up there?" gesturing toward the waterfall. She gives him a puzzled look and indirectly answers his question, saying, "My friend went climbing up there." We explain that we received a 911 phone call from someone who had fallen near here. The entire conversation is difficult, both because of the helicopter noise bouncing off the steep rock walls and because the woman doesn't know why rescuers are suddenly arriving en masse.

As we are talking, the pilot calls on the radio to say that they have spotted the victim on a ledge near the top of the waterfall. Leaves settle from the sky as the helicopter banks sharply and flies away, taking its portable windstorm.

A "Lucky" Fall

Working our way around the left side of the waterfall, Gary, Dan, and I have to climb higher than the falls to find a safe route across the steep rock wall. While traversing the wall, we look down and see the victim on a ledge in the waterfall. He is lying on his back in a shallow pool of water, limbs splayed out, not moving. A lucky thing, landing on the ledge, since he is still seventy feet above the deck. He managed to reach his cell phone and call 911. The victim will later recount that when he saw us above him, he told the 911 operator, "I see them!" and that "the dispatch lady cried."

At the top of the waterfall, several rescuers set up an anchor on a nearby tree and throw down the rope, which the current quickly drags over the falls. With the helicopter gone, we can hear the victim's primal screams of pain.

Gary rappels down the waterfall, taking care not to slip on the greasy rocks. The victim, eighteen-year-old Travis Metzger, is lying in several inches of cold water. The rocky shelf is slimy with bright-green algae, and I'm a little surprised that the shallow moving water hasn't slid him over the edge to his death.

Gary immediately sets up anchors by inserting spring-loaded camming devices, whose metal lobes expand outward when weighted, into cracks in the rock, and then secures himself and the victim to these anchors.

Dan rappels down next and, after securing himself to Gary's anchors, releases the rappel line so I can rap down next. Dan has the team at the top of the falls lower the full-body vacuum splint and the orange litter, while I begin a medical assessment of the patient. I quickly discover that although he has some feeling in both legs, he can't move either. His inability to move his lower extremities, even wiggle his toes, points to a serious spinal cord injury. He has cuts on both arms, and his skin is pale, cool, and clammy, either from shock or from being in cold water for the past hour.

As I talk to Travis, he tells me he clearly remembers falling and that he did not lose consciousness. He says he first fell twenty to thirty feet onto a rock ledge just above our location. He then fell the remaining eight feet onto this shelf. He knows he is cold, paralyzed, and on the edge of a waterfall. I can see the pain and fear in his eyes.

We put a cervical, or "C," collar around his neck. The collar minimizes movement of the head and neck, and—perhaps as important—it serves as a constant reminder to the patient not to move.

Using foam-covered flexible sheets of aluminum known as SAM splints, I stabilize both his ankles. We then secure the full-body vacuum splint in the litter. The vacuum splint is a bright-red rectangular air mattress made out of heavy-duty vinyl. It contains thousands of little Styrofoam pellets, which is why we often call it simply the "beanbag." We roll the sides of the beanbag around the patient and secure its eight wide straps. Then, as we manually pump the air out of the bag, the pellets are squeezed together, forming a rigid splint that conforms to the patient's body contours. Once formed, the beanbag immobilizes the spine much better than traditional backboards, which can allow the patient to slide around while jostling and tilting over mountainous terrain. The conforming beanbag is also much more comfortable than the hard, flat backboard—not that important on a quick ambulance ride from house to hospital, but it makes a huge difference if you're strapped in for a bumpy multihour trip down a mountain trail.

We call this process "spinal immobilization," because we want to prevent movement that might cause fractured vertebrae to sever spinal nerves, but neither a backboard nor a vacuum splint can completely prevent movement. A more appropriate term is "spinal stabilization," since we can only reduce, not eliminate, spinal movement. Travis shows clear signs and symptoms of a spinal cord injury, and we're doing our best to keep it from getting worse.

We attach the litter to a spider bridle: four thin ropes connecting to the sides of the litter and to the ropes from above. I clip an adjustable length of rope from my harness to the bridle.

Two more teams of rescuers, led by Tom Moyer and Darren Hunsaker, have been rigging lowering systems high above us on the rock wall we traversed. With a team also at the top of the waterfall, Travis and I are connected to three ropes.

Gary, Dan, and I lift the loaded litter as the three teams take the slack out of their ropes. We can't simply lower him over the ledge, or the water from the falls will pour onto the horizontally packaged patient and me, possibly drowning him. Instead, we need to transfer our weight from the rope at the top of the waterfall to the two ropes to our left, shifting us away from the waterfall and over to the rock wall.

Safety in Redundancy

Being lowered with a patient on a litter is not very different from rappelling. Although it may look impressive on TV, I'm probably safer during this rescue than I was driving here. Three separate teams of rescuers are lowering me on three ropes, each of which can hold more than 6,000 pounds. That's enough strength to support three Lincoln Navigators.

Rope rescuers always strive for redundancy. In fact, many teams require complete redundancy by imposing the "scissors test"—which insists that if any component of the system is cut away by an imaginary pair of scissors, the entire system must not catastrophically fail. That level of redundancy isn't present in most of our daily activities—our cars have just the one brake pedal, for instance, and one steering wheel—but it's a good objective in rope rescue.

As I struggle to pull the litter away from the waterfall, the three teams of rescuers are busily adjusting the tension on the ropes, to swing us away from the flow. Their voices resonate from the radio on my chest. Knowing that I should be safer right now than while driving a car helps calm my nerves—a little. But the reassurance really comes from knowing I can trust my teammates on the other end of these ropes. From here on down, I'm just the dope-on-a-rope.

I talk casually to Travis as we descend the rock face, acknowledging that I bet he didn't expect to be lowered down a rock wall today. I'm hoping my relaxed tone will help him feel a little better about dangling on a rock wall, at the mercy of utter strangers.

When we reach the ground, the crew from Life Flight takes over patient care. As always, I feel a sense of relief in passing off our seriously injured patient. Other rescuers then pick up the litter and carry it down the trail to the waiting helicopter. I will later learn that doctors spent six hours performing surgery on Travis' crushed spine.

Throughout the rescue, I noticed two or three tiny specs in the sky. They were helicopters hovering a mile or more away. I knew they might be filming us, but I had no idea they would be shooting close-up pictures throughout the rescue or that the whole thing would be running live on several TV stations. One of the stations even bumped its regular news broadcast, and the two anchors, the meteorologist, and even the sports anchor narrated the rescue as it happened. They even brought in an experienced technical rescuer from the Salt Lake City Fire Department to help with the blow-by-blow coverage. Travis's mother, who was living in California, watched the rescue live on the Internet.

Steve and the patient prepare to be lowered
(Visit MountainResponder.com/Photos and enter Story Code RMW.)

The Media: Good, Bad, and Inexcusable

The media can be a great asset to rescuers. They help the public see what we do, or at least a little sliver of it, and this increased public awareness no doubt causes some people to reflect on how their actions might put themselves or others in unnecessary danger. The media were also helpful when we wanted to get a message out to the public, such as a warning about increased avalanche hazard or the danger of swiftwater during spring melt. They also helped us locate missing people when we weren't sure if the person was truly lost.

The flip side to this helpfulness is that reporters are in the business of getting stories—and getting them fast. After I became commander of the team, it was typical for my phone to ring a few minutes into a callout. It was usually a deputy providing additional information. But too often it would be a reporter saying something like, "Hi, this is Joe with KSL Radio. What can you tell me about the avalanche in Big Cottonwood?" I learned to reply immediately, "I'm busy," and hang up. I also learned to save the number on my cell phone as "Media," so I'd be forewarned the next time they called. I understood that these guys were doing their job, but I was doing mine.

On rare occasions, a reporter would cross the line. After recovering a body in Bells Canyon, I asked the team to "circle up" for our debriefing. A reporter with a high-tech tape recorder slung over his shoulder joined our circle. I pulled him aside and explained that the point of a debriefing was so rescuers could speak freely to each other about what went right and what could be improved. Assuming that my explanation would help the reporter understand and respect our need for privacy, I returned to the circle to continue the debriefing. Sure enough, there he was, lurking behind two rescuers, with his back to us but with his microphone slipped between them. Interrupting the debriefing, I crossed the circle, parted the two rescuers, and turned the sleazy newshound around. And in language rich with expletives and colorful imagery, I explained to him the immediate risk to his recording equipment should he linger another moment anywhere near our little circle.

Four
Falling Climbers

When people think about mountain rescues in Salt Lake County, they usually think about rock climbers. Considering the world-class climbing in the Wasatch range, that's a reasonable hunch, but in fact, fewer than one in ten calls involved rock climbers. That is roughly the same proportion of callouts that involved backcountry skiers and snowboarders, but it is dwarfed by "hikers," who accounted for almost half our calls. Granted, "hikers" includes people (often inexperienced) climbing without ropes—like the young man who fell while climbing near the Rocky Mouth waterfall—because we classify only people who are climbing with ropes as "rock climbers." The rest of the callouts resulted from a broad range of activities, each of which accounted for less than 3 percent of our calls. These included such pursuits as horseback riding, mountain biking, hunting, swimming, hang-gliding, driving ATVs or snowmobiles, and aviation.

Now, it's worth noting that because much of the rock climbing terrain in Salt Lake County occurs within a half-hour's hike of the road, many injured climbers no doubt managed to hobble down to their cars with the help of friends and never called 911.

As in other activities, where people were doing a wide variety of things when they got into trouble and needed rescuing, the same was true for rock climbers.

Pentapitch

I'm sitting at my daughter's soccer game on a cool, clear afternoon and trying to figure out what that strange, faintly annoying beeping sound is, when I realize—*blush*—it's my pager. Having been on the team only a few weeks, I have yet to form an inseparable bond with the pager.

```
CALLOUT, FALLEN HIKER, LITTLE CTWD
CYN, RESPOND ON CODE-1 CHANNEL. CS
54947. (EOM)
13:41 05/19/01
```

Since this is only my third callout, I take the words "fallen hiker" literally. I will later learn that the term can refer to somebody who slipped and cut her knee in a picnic area, a kid who fell off a rope swing, or a rock climber who has fallen to his death.

The pager tells us to go to Little Cottonwood Canyon, but it doesn't say where. I have the newbie's concern that I'll screw up, but I decide to drive up the canyon and look for a cluster of emergency vehicles. I'm not sure what I'll do if I get all the way to Alta Ski Resort without seeing my companions, but reaching the mouth of the canyon, I realize that my concern was unfounded. Traffic is stopped near the mouth of the canyon, and a deputy is preventing more cars from entering. The driver in front of me turns on her emergency flashers and slowly pulls into the oncoming lane. It's my training officer, Amy Fisher. I follow her up the twisting canyon road.

Parking behind Amy, I strap my radio onto my chest and pull on my hiking boots. Following Amy's cue, I also stuff my climbing harness into my pack. As I'm doing this, a helicopter lands on the road about a hundred feet from us. Aware of the spectators in the queued-up cars, I'm feeling undeservedly important.

Amy and I gather our medical gear and head up the trail. Bystanders who have posted themselves at the two trail junctions direct us toward the victim.

The climber is a twenty-four-year-old male who, after setting several pieces of protection on the second pitch of "Pentapitch," was having a hard time pulling the rope up behind him. Looking down, he saw a sharp bend in the rope where it doglegged around an angular rock. He began downclimbing to his last piece so he could add a runner (loop of nylon webbing) to lessen the kink and allow the rope to run more freely. When he was ten feet above the protection, he fell. This resulted in a 20-foot fall—the ten feet down to his last protection, and the additional ten feet that he must fall before the slack in the rope pulled taut. When the rope came taut, his protection failed to hold. Thus, he now had twenty feet of additional slack as he plummeted toward his next piece— which also failed. Continuing his horrifying descent, he shot past his third piece of protection, a sling around a tree. It held.

The fall happened on steep, though not vertical, rock, resulting in a cheese-grater effect as the crystal-embedded granite slowed his descent.

After his forty-foot fall, the climber was dangling a few feet below his belayer. Covered with blood and small cuts, he had a broken tooth, lower back pain, and a concussion—he was not wearing a helmet. His two friends hauled him up to a nearby ledge and then lowered him down the first pitch. He was preparing to walk out when the first rescuers arrived.

When Amy and I arrive, our teammates are already packaging the bloodied climber in the full-body vacuum splint— the beanbag—and litter. They connect a rope to the litter, and we begin carrying him over desk-size granite boulders. We work like an army of ants, taking turns scurrying ahead of the next obstacle so we can keep our precious bread crumb moving along. When we reach the end of the rope, I see that a forward-thinking rescuer is already waiting for us with a second anchor. We reconnect our rope and continue downward.

When we reach the road, we load our patient into the waiting ambulance. I glance at my watch and see that it's only been an hour and a half since my pager went off—since it's one of my first rescues, I don't yet appreciate just how rare a thing it is to complete a mission this quickly.

I reflect on the obvious irony that this climber was injured while trying to improve his protection. I'll see something similar tomorrow.

Aqueduct Wall

I'm at the climbing gym with my ten-year-old daughter and her friend when the call for another fallen hiker appears on my pager. It was just yesterday that we brought down the climber from Pentapitch. The text on my pager directs me to another popular climbing area.

An experienced climber has fallen while leading Sockdollager, a 5.9 climb on Aqueduct Wall. *Sockdollager?* The creativity—and sometimes the lack of it—involved in naming these climbs is an endless source of amazement and amusement. The five-rope-length Pentapitch is logical, but I wonder about the story behind Sockdollager.

About ten feet up the hand crack, the twenty-five-year-old placed his first piece of protection. After climbing ten more feet, he inserted a second cam into the crack, but as he reached down for the rope that was trailing below him, and pulled it up to clip it into the protection, he fell. Granite shards flew as the first piece blew out of the rock. After dropping twenty-five feet through the air, he smacked into the rocks several feet below the block where his belayer was standing. A climber on a nearby route, who happened to be an EMT, witnessed the fall and came to the injured climber's aid.

As with the climber on Pentapitch, this climber took a bad fall while trying to protect himself from injury. And, like the other climber, he wasn't wearing a helmet.

It's a short hike from the road to the victim's location. When I arrive, I see that the climber came to rest in a cavelike hollow below a bus-size boulder. Commander Gary Banks is stooped under the rock, straddling the patient, and providing medical care. Someone has already removed the high-tech climbing shoes, and the protruding feet are gray as paint, like something sticking out of a morgue drawer and wearing a toe tag.

The climber has head lacerations and is complaining of back pain. After putting him on oxygen, Gary gives the familiar EMS commands: "Lift on three… One, two, three!" The goal of our coordinated lift is to minimize movement of his spine, but the little cavern is cramped, and the rescuers have to make do. As I watch the rescuers lift the patient and slide the litter and vacuum splint under him, I hear the stern schoolteacher within me saying, "This isn't how they taught us in EMT school." I will learn that less-than-ideal field conditions, from rock recesses to cliff bands to avalanche hazards, occasionally force us to deviate from protocols in the care we provide.

After securing the patient in the litter, rescuers lower him down several pitches of steep, loose rock using dual ropes. Gary has us "newbies" stand to the side during the lowering. I don't like being treated like a fifth wheel, but years later, I'll find myself in Gary's role, making similar decisions. My, how different things look depending on the hat we're wearing at the time!

We transfer our patient to the white-and-orange ambulance. They drive him a mile down the canyon where he is transferred to the waiting Airmed helicopter and flown to the hospital in critical condition.

A few days later, I see this young man on the ten o'clock news, with a steel "halo" around his head, connected to his skull with metal pins and to a rigid vest with vertical bars. This Rube Goldberg contraption supports his head and limits motion of his broken neck. Our patient is smiling happily as he talks to the reporters. In addition to a fractured cervical vertebra, he has two fractured vertebrae in his lower back, a bruised kidney, broken ribs, a sprained ankle, and a concussion. Although my care extended only as far as helping secure him in the vacuum splint and litter, it feels good to see the positive results.

Perhaps

It's early in the evening and the late November days are shortening as I sit in front of the TV. My adrenaline delivery device, cleverly disguised as a pager, delivers its now familiar jolt.

```
S/R CALLOUT FOR FALLEN HIKER. CARS TO
MEET AT GATE BUTRESS. THIS IS ON CD1
FREQ. CS# 140241. EOM.
20:34 11/19/02
```

A group of three men were climbing at dusk. The leader was on the second pitch of Perhaps, on the Gate Buttress. This pitch ends with a long horizontal traverse. The climber placed a cam at the beginning of the traverse and clipped an old piton about fifteen feet out. After traveling another fifteen feet, he stopped and looked up for the final anchor chains. As he did, his feet skidded out from under him, and he fell. After dropping twenty feet, his plummeting weight yanked the rope taut on the "fixed" piton, which popped out of the rock, and he continued his sixty-foot pendulum arc. The cam he set at the start of the traverse held as he slammed into a small ledge.

Unlike the climbers on Pentapitch and Sockdollager, he didn't get his injuries while trying to improve his protection. It was his reliance on the decades-old piton that cost him. The steel pin was hammered into the rock back in the days when this was environmentally acceptable, in a spot where traditional protection is not to be had. In hindsight, inserting pro as soon as possible after passing the piton would have been a good idea, but we all clip the occasional fixed pin that we come across, often without a second thought.

The fallen climber's two partners lowered him to the top of the first pitch, rerigged, and then lowered him to the base of the climb—a good start to a self-rescue.

When we arrive, I perform a medical assessment while two other rescuers build anchors and assemble the litter so we can lower the patient down the long, steep talus field below us. Inspecting his ankle, I see that the foot is pale blue and the ankle is deformed, with the foot rolled inward in an unnatural position. It's pretty obvious that the ankle is fractured, dislocated, or both. I check for a pulse on the top of his foot—the dorsalis pedis artery—and can't feel anything. I squeeze his big toe, pressing the blood out from behind the toenail, then release it and count to three. His nail bed remains pale. This informal test measures how long blood takes to flow back into the small capillaries behind the toenail. The absence of a pulse and of capillary refill are ominous signs, since, without fresh oxygen flowing to the foot, it will die. But unlike heart muscle or brain tissue, which begin dying almost immediately, the death of the foot will happen slowly, over hours.

Before splinting a fracture, the limb or extremity should be returned to the "anatomically correct" position. This is especially critical if the limb is pulseless, like this injured foot. The realignment procedure calls for gentle inline traction to return the limb to its natural position. Holding the heel and foot in my gloved hands, and with Gary stabilizing the patient's lower leg, I give a slow, steady pull to restore it from its inverted position. Rather than getting a restoration of circulation, I get a blood-curdling scream.

Rob Hunter, a skilled paramedic who worked as a mountain rescuer in Yosemite National Park, is the senior medical rescuer on scene. I tell Rob what I tried, and the results: still no circulation. We agree that I should splint the ankle in its current position. When I ask Rob if we should have an air ambulance waiting at the road, he says, "We don't need to decide that yet." He's right—we're still hours from the road.

We put him on oxygen, splint his foot, package him in the beanbag splint and litter, and begin lowering him down the six pitches of steep talus. Each time we stop to connect the rope to a new anchor, we recheck his vitals and the circulation in his foot. On the third pitch, Rob notices signs of circulation. Not an arterial pulse, but enough circulation to see capillary refill and a slight pinking of the foot.

When we reach the road, we quickly transfer the patient to a waiting ambulance. The next day I hear that he had fractures of both lower legs and both ankles, and five compression fractures of his lower spine. This is consistent with the injuries of other fallen climbers I've seen. They tended to land catlike on their feet, though perhaps not as gracefully. They break bones in their lower legs and feet, have compression fractures of the lower spine, and often have a head injury. The spinal fractures remind me that the injured climber and his two friends were embarking on a self-rescue when we arrived—walking out would have been a very bad idea.

Considering the length of his fall and the deep scars in his helmet, he was lucky he didn't have a serious head injury. Had he not been wearing a helmet, things would have gone very differently.

Steve attempts to restore circulation to the patient's foot
(Visit MountainResponder.com/Photos and enter Story Code PHC.)

Hollow Man

Of course, not all our climber rescues were the result of falls—occasionally, a simple mistake triggered a rescue situation.

It has been less than two hours since our last call—an injured backcountry skier—when my pager/adrenaline pump fires up.

```
This is a callout for S/R on a
stranded hiker...meet car 614 at
Storm Mountain in Big Cottonwood Cyn..
technical climber gear needed...
respond on Code one Freq..case39502..
eom
04/18/04
```

A twenty-five-year-old man had just led Hollow Man, a sport route on Challenge Buttress. Most climbs in Salt Lake County are "trad," that is, traditional, which means the climber must insert protection into cracks or pockets in the rock. But there are also many "sport" routes, which have "hangers"—sturdy metal tabs with a hole to accept a carabiner—bolted to the rock. The climber simply clips into the hole in the hanger and clips his rope in as he passes. The advantage of a sport route is that clipping a bolt hanger is a lot faster than setting pro on a trad route, and the bolts are almost certain to hold a falling climber—unlike trad gear and aging pitons, both of which occasionally fail.

At the top of a sport route (and of some trad routes as well), two short lengths of chain or specially constructed metal rings are often bolted to the rock. Reaching the top, the climber clips carabiners to the chain ends or rings, passes the rope through the carabiners, and then rappels, or is lowered, to the ground. Then another climber in the party can climb the pitch while being belayed from below.

In this instance, on reaching the top, the climber clipped his harness into the bolts and untied the rope from his harness. All very standard procedure—or it would have been had he thought to tie a hitch in the rope and clip it to a carabiner so that it couldn't get away from him and fall to the ground if he dropped it. And this thought no doubt occurred to him when, as he started to thread the rope through the carabiners, it did indeed slip out of his hands and he watched it snake its way to the ground, leaving him stranded on the wall.

His climbing partner was quite capable of leading Hollow Man and bringing the rope back up so the stranded partner could rappel or be lowered off, but alas, she had no one to belay her. Using her cell phone, she played "dialing for belayers," calling every climbing friend she could think of, with no luck. As the sun began to set and the April air began to chill, she dialed the one number where she was sure to get an answer: 911.

When we arrive, the belayer is quite vocal in telling us how we should proceed with the rescue, but Gary does a fine job of ignoring her. I belay a rescuer who leads Hollow Man, while Gary and another teammate climb up and around to access the chagrined stuck climber from above. He is lowered safely to the ground.

After several hours of hanging in his harness, the guy is cold and uncomfortable, not to mention a bit sheepish at having put himself in such an awkward position. For me, though, it's just refreshing to rescue a climber who hasn't suffered anything worse than a bruised ego.

Penitentiary Wall

Excluding the unusual case where someone dropped the rope and needed to be rescued, most climbers were on the ground when we arrived. That's either because gravity did its job and deposited them at the base of the climb, or because the partner was able to lower the injured climber. Usually.

```
This is a callout for S&R for a
fallen hiker..meet at Ledgemere
picnic area approx 4600 e
BigCottonwood..respond on Tac One
Frequency..case#45574..eom
12:52 05/26/06
```

As I'm driving to the popular climbing area, radio updates inform us that a climber has fallen, is unconscious, and is still hanging by his rope.

Parking on the side of the road, I see several firefighters heading up the trail with their bulky full-body harnesses and thick, unwieldy urban rescue ropes. I see their squad leader—a captain, I suppose—hand a rigid backboard to one of his men.

I introduce myself and I ask him what gear has already gone in. I also suggest that we'll probably want to use the full-body vacuum splint rather than the backboard.

Towering over me and outweighing me by a good seventy pounds, he informs me that we'll be using only ropes approved by the National Fire Protection Agency and not "any of that climbing gear." With a critically injured climber dangling in dire need of rescue, he starts reciting the rigging standards that must be met on this call, insisting that "everything must pass the scissors test" and that they will not be using our full-body vacuum splint.

I'm stunned that this guy has appointed himself the rope rescue czar, that he's sending in completely inappropriate gear for a mountain rescue, and that he's making these unilateral decisions while a climber is hurt and perhaps dying. And along with their having all the wrong equipment, I'm willing to bet that his men have no climbing experience—zilch. I can taste my frustration and anger. What in the hell is this guy *thinking*?

Putting on my best "Don't make me get rough" game face, I take a deep breath, calm my quaking knees, and push my finger into his chest. I tell him, "By state law, the sheriff's office—*not* the fire department—is responsible for search and rescue missions. So back off!"

I wonder if he can hear my voice tremble. He says something about me being full of crap as my teammates and I pick up our mountain rescue gear and head up the trail. I'm seething. I can't believe some guy who may have a ten-day rope rescue course under his belt is telling us how to rescue a climber! I actually had to play the "state law" card! Does he really think his men are going to get a dangling climber down safer and faster than people who have done it many dozens of times?

As I scramble up the loose scree toward the climbing area, I know I have to let go of my frustration. I try to let the strain of climbing up the slope with a heavy load pump the exasperation out of my system. I'm glad the guy is staying at the trailhead—the rank-and-file firefighters ahead of me, who know their strengths and limitations, will be much easier to work with.

After toiling fifteen minutes to get to the base of the climb, I see a climber half hanging, half perched on a two-foot-wide flake of rock, fifteen feet up Penitentiary Wall. His rope is still tied into his harness and appears to be tangled above him in a way that prevents him from falling the rest of the way to the deck.

One of the first things a rescuer does on reaching an injured person is to ask what happened. It just helps to know whether they slid or plummeted, bumped or slammed, fell five feet or a full rope length. The answers give insight into what's known in medical jargon as the "mechanism of injury." The mechanism can figure heavily into treatment and transportation decisions.

Immediately after this fall, a friend of the victim, who was on an adjacent climb, was able to climb over to the victim as he hung near a rock ledge. The belayer ran to the road to call for help and told a deputy that the rope "just started coming down all at once."

When we ask bystanders what happened, they tell us, "He fell fifteen to twenty-five feet," and that "he was knocked out for about ten minutes." Looking at the damage to his helmet, I wonder whether he could have survived without it.

There are several possible causes of his fall, and I'm curious, but our immediate challenge is how to get him safely down from where he's hanging.

Three rescuers—a firefighter, Alan Bergstrom, and I—climb the fifteen feet to the small ledge. Using several cams, Alan builds an anchor in the wall above us, while the firefighter struggles to put a cervical collar on the patient, who is combative and is pushing him away.

Using a Prusik, I tie myself into the climber's rope to limit my chances of falling. I say "limit" because although the rope is holding the victim up, I can't be sure of its reliability—it is probably through only a single bolt, has already taken a hard impact, and may be damaged as well. But with the four of us perched precariously on the narrow shelf, I welcome even this questionable protection.

Meanwhile, rescuers on the ground assemble the litter and vacuum splint and toss us a rope, which we thread through Alan's anchor and pass down to the rescuers. Pulling on the rope, they haul the litter up the wall until we can slide it under the struggling climber.

With the litter behind him, we attempt to fasten the vacuum splint's straps around his legs, but he's kicking and pushing us away. Combativeness like this is a typical symptom of a head injury. Add to this that his eyes open randomly and roll about without focusing, he doesn't respond to our instructions, and he's making babbling sounds but no actual words, and it's pretty clear we have a guy in serious straits. After several minutes of struggling with him to immobilize his spine, we decide that he needs to get to a hospital more than he needs textbook spinal immobilization. Grasping his arms and legs, we strap him down in whatever position we can. This is one of the few times we will transport a patient before stabilizing him.

With the patient bound to the litter, rescuers lower him to the ground. A crew of six rescuers and firefighters then lower him down the steep scree, using tree wraps for brakes. Although I'm still frustrated that I had to get in the fire boss's face at the trailhead, the firefighters who actually went on the mountain were a great bunch to work with.

When we reach the road, a helicopter is waiting to take the patient to get the medical care he desperately needs. Later I see on the news that he is being kept in a medically induced coma, a sort of "hibernation" to let the brain recover, as they monitor the seriousness of his head injury.

Turf Wars

The firefighters we worked with on the Penitentiary Wall rescue were top-notch—professionals who understood teamwork. But, as we saw in their boss's earlier attitude at the trailhead, when Fire and Rescue are on the same call, there's often a certain tension at work. Usually it's an undercurrent, but sometimes it flares up right out in the open. Some firefighters squirm at seeing volunteer rescuers going in. They see themselves as the ones who should be doing *all* rescues, even in the backcountry. Likewise, many mountain rescuers—and this definitely includes me—get uncomfortable when we see firefighters heading into the mountains.

It isn't that I question firefighters' medical skills—most of them are very solid in this area—but most have little or no real backcountry rescue experience. Some assume that because they have spent time camping, hunting, or fishing in the great outdoors, this experience will somehow combine with their urban rescue skills to make them equally competent in the mountains. Alas, the reality is not so simple. This overestimation of one's own outdoor rescue skills, and the assumption that city skills will readily apply to backcountry situations, is half the problem. The other half (and this is from my admittedly biased perspective) stems from many firefighters' failure to understand that most mountain rescuers have been doing some combination of climbing, hiking, paddling, and backcountry skiing for years, if not decades.

The same combination of inexperience and misperception also occurs with some law enforcement personnel. The cops I know are consummate professionals when dealing with criminals or keeping public order, but very few of them have any backcountry or rescue experience.

I don't think most firefighters or law enforcement officers should participate in complicated backcountry rescues, for the same reason I don't want my accountant taking my appendix out—it isn't his area of expertise. In this same vein, a backcountry rescuer is not the guy—or gal—to get you out of a burning building or a crumpled car, or to stop a rampaging sicko with a gun. Whatever the job, it tends to come out better when it's done by the people who are trained to do it. And therein lies the problem.

Firefighters and law enforcement officers can certainly learn the basic avalanche, climbing, rope rescue, swiftwater, and other requisite skills to become mountain rescuers. I know this because I've had the pleasure of working with some excellent ones. And those who really get it know that a weekend course or two won't do the job—to keep up and hone and improve their mountain skills, they'll have to spend years in the backcountry.

I've been on rescues where the responding agencies worked smoothly together, and I've been on some where we butted heads. On the best calls, all the participants recognize their roles, strengths, and weaknesses. The mountain rescuers locate, access, stabilize, and transport the victims to the road or landing zone. As they are doing this, they communicate with the urban medical responders, who can provide additional medical advice over the radio. When the rescuers get the victim to the road or the LZ, the paramedics take over, providing advanced medical care and transporting the patient to the hospital. At the emergency room, doctors take over and provide a yet higher level of care. If required, the patient is transferred to surgeons or other medical specialists for further treatment. As the patient recovers, physical therapists help to restore pre-accident function. Working together, these various professionals do what they are best at, with each committed to the best possible outcome for the patient.

When it doesn't work, it's the patient who suffers. I remember skiing in a snowstorm, pulling a toboggan with a critically injured patient. She had hit a tree, was fading in and out of consciousness, and barely had a pulse, and we were working hard to get her out of the mountains alive. As we skied by two firefighters who were trudging uphill in their yellow canvas turnout pants, I again reminded the incident commander that we wanted advanced life support ready and waiting at the road. I knew that our patient needed care beyond our abilities. The command post said that paramedics were standing by in a nearby ambulance. When we reached the snow-covered road, someone handed me a towrope connected to a snowmobile, and the driver towed us several hundred feet up the road, to the waiting ambulance. Swinging open its rear doors, I was surprised to see two EMTs and no paramedics. The EMTs looked as surprised as I was. It turned out that the two firefighters we passed on the mountain were the paramedics. The EMTs had few options but to put the patient on oxygen (ours had run out), warm her in the ambulance, start an IV, and wait anxiously while the paramedics hurried back down the mountain. The patient turned out to have a fractured pelvis and internal bleeding. The outcome was good—she lived, although she spent months recovering in a wheelchair.

And there was the time the urban emergency medical service responders put a fallen climber on a slippery plastic backboard rather than using a full-body vacuum splint. As they were lowering him down a steep slope, the patient slid forward on the backboard, and the straps that were intended to hold him on the board began to strangle him. A well-meaning EMT gave a big push on his feet to slide him back on the board. That was the first time I heard the otherwise unresponsive patient respond—with a loud, agonized moan. He had a fractured neck.

But the antisynergy grand award surely goes to that Fire boss on the Penitentiary Wall rescue, who insisted on quoting chapter and verse from the National Fire Protection Agency while a guy was up on the wall, unresponsive and maybe dying. There comes a time when you have to swallow a little agency pride and do what's best for the victim.

Without doubt, most firefighters and law enforcement officers are great rescue partners—knowledgeable, hardworking, and professional enough to distinguish their own skill sets from someone else's. But I suppose there will always be that arrogant minority who don't recognize the difference between urban EMS and mountain rescue and who don't seem to understand that the best and safest outcomes result from focusing on our core competencies—not from trying to overlap into a job that someone else is better trained for.

Five
Tumbling Snow

In the United States, some twenty-five people are killed by avalanches each year. To people who haven't witnessed an avalanche, it may be difficult to believe that moving snow can kill you. After all, as we see it falling in the movies, snow has the consistency of fluffy down feathers, right? Wrong. According to Bruce Tremper, author of *Staying Alive in Avalanche Terrain,* "Avalanche debris seizes up like concrete the instant it comes to a stop."

I experienced the "concrete effect" firsthand when I was searching an avalanche for buried cars. As I swung my steel shovel with all my might into a large block of avalanche debris, the sharp edge bounced off the snow leaving only a half-inch nick. Get under even a foot or two of that that stuff, and you have a serious situation.

A deputy arrives at the massive avalanche
(Visit MountainResponder.com/Photos and enter Story Code LCA.)

Some avalanche survival stories say that if you are buried, you should spit so gravity will show you which direction is down, and then dig in the opposite direction. Even if you are fortunate enough not to have snow packed so tightly into your throat that you can't possibly breathe, knowing which direction is up still won't solve your dilemma—you will be so encased in this concretelike material that you will barely be able to wiggle a finger, let alone dig in any direction. The best thing you can do to survive an avalanche is to avoid it in the first place—if you truly wish to avoid being buried, call your local avalanche forecast center *before* venturing into the backcountry.

Getting Found

If you are buried by an avalanche, survival hinges on two factors: time and trauma. In the United States, about a quarter of avalanche victims are killed by traumatic injuries as they crash into trees or are hurled over rocks and cliffs. And the chance of survival quickly plummets the longer the victim is buried. Of those who survive the ride, more than 90 percent of victims will make it if they are extracted within fifteen minutes; after that, the numbers drop rapidly. Only 40 percent survive if they are found after thirty minutes, 28 percent survive for one hour, and only 10 percent survive after being buried for two hours. Only 2 percent live long enough to die from hypothermia.

Surface Find

Most people who are caught in avalanches aren't buried and can ski away, frazzled. Sometimes, though, we are called to avalanches where the victim, though not buried, can't ski down, because of injuries or missing skis. And, of course, staying unburied is not the same as staying alive.

```
Callout - missing skiier 9300 e Big
Cottonwood Canyon stage at Butler
Fork Bring Winter Gear and come 10-41
on TAC-2 channel. case#07-14991 eom
19:40 02/21/07
```

When we arrive at the command post, we are told that a visiting Norwegian man was backcountry skiing with two friends when he decided to break off and take a few runs on his own. He didn't return at day's end.

The avalanche hazard today is "considerable." In the well-defined jargon of avalanche forecasting, this means, "Avalanches are probable with human and natural triggers." Today's avalanche report was even more ominous:

The avalanche danger remains CONSIDERABLE on slopes steeper than about 35 degrees facing northwest through northeast through southeast, where dangerous avalanches 1 to 3 feet deep can be triggered by people. Back off the steep stuff – if the close calls continue, someone else is going to get killed or hurt.

Given the terrain, which was northeast facing, and a report like that, we decide that the hazard is simply too high, and the search area too large, to send in rescuers at night. Instead, rescuer Tom Moyer goes up in a helicopter to see if he can spot our victim from the air with night vision goggles. Tom is a ski patroller, an engineer with degrees from Stanford and MIT, and our team's undisputed rope rescue wizard. He finds an obvious avalanche, with "tracks in but not out." There is no sign of the missing skier.

The next day, Tom and I are flown in with three people from Wasatch Backcountry Rescue, a winter rescue team of ski patrollers from local ski areas. We eventually find the victim on top of the snow, with his face pressed into a tree. Though not buried, he was killed by trauma as the avalanche strained him through a stand of tall pines.

Steve and rescuers transport the avalanche victim
(Visit MountainResponder.com/Photos and enter Story Code GKA.)

Sometimes, though the victim is buried, there is a visual clue to his location. On one of our calls, rescuers found a hand protruding from the snow. It was attached to a dead snowboarder.

And then there was the call on April 30, 2005, where a father and son, aged seventy-eight and thirty-seven, were "poking" at the snow cornice hanging from the roof of their cabin. The snow sloughed off the roof and buried both men. Family members inside the house heard the slide and ran outside to find a pile of snow eight to ten feet high. A kicking leg was protruding from the pile. They ran out to the street and flagged down a neighbor who was plowing snow. Someone called 911, and the Search and Rescue Team was paged. Several deputies were on scene within minutes of the call with shovels and avalanche probes.

The thirty-seven-year-old man, who was under two feet of snow, was unburied within five minutes. His breathing was shallow, and he was cyanotic (bluish-purple).

Within another ten minutes, the rescuers unburied the arm of the seventy-eight-year-old man. Using shovels, they followed his arm to his head. He, too, was cyanotic and barely breathing. Rescuers continued to dig out his legs, which were buried under four feet of snow and large blocks of ice.

Due to low clouds that prevented air ambulances from responding, both patients were transported down the canyon in ground ambulances; the older patient was transferred to an Airmed helicopter at the mouth of the canyon and flown to the hospital. Roof avalanches are a very real hazard and have killed more unsuspecting people than one might suppose.

This roof avalanche nearly killed two men
(Visit MountainResponder.com/Photos and enter Story Code BRA.)

Photograph by Darren Hunsaker

Transceiver Searches

If you end up below the surface, your best chance of being found is with an avalanche transceiver. Avalanche transceivers transmit a silent beep every second or two. If you are buried, your partners switch their units to receive mode and use them to find you.

Efficient searching with a transceiver requires practice. Most rescuers and patrollers practice using their transceivers at least monthly—waiting until your best friend is on borrowed time isn't a good time to learn the intricacies of your electronics. The unfortunate Norwegian was wearing a transceiver, but it broke on impact with the tree. Of course, he was skiing alone, which would have rendered the device useless anyway, since there was no one to search for him.

To learn more about avalanche survival,
visit BeaconReviews.com

If you are lucky enough to be found with an avalanche transceiver and live to talk about it, it will almost certainly be because your partners, not professional rescuers, found you. With only 10 percent of avalanche victims surviving a two-hour burial, it's unlikely that your friends can call 911 and have rescuers on scene fast enough to save your bacon—unless, like the lucky duo in Brighton, you happen to get buried right outside your home, and someone notices immediately.

This call, too, involved some unusual good fortune:

```
SEARCH AND RESCUE CALL OUT. AVALANCHE
VICTIM. CARDIFF FORK BIG CTNWD CYN.
RESPOND ON CODE 1. SNOWCAT AND
SNOWMOBILE NEEDED. CASE 2035.EOM
13:56 01/06/03.
```

A man was backcountry skiing alone, in "considerable" avalanche danger—note that he's already committed two grave errors—when he triggered an avalanche and was buried. Miraculously, two skiers witnessed the slide, skied down, located him with their transceivers, and had his head out within "ten to twelve minutes." It was surely this guy's lucky day, because the helpful skiers were a doctor and a nurse. And his streak of luck continued when two ski patrollers, out for some backcountry skiing, happened to look down into the bowl they were about to ski. Seeing the shoveling, they radioed for a helicopter with rescuers and an avalanche dog. This soloist survived with only minor injuries.

Although I hope my transceiver will help me locate a live victim, it hasn't happened yet. And while it would even be good if my transceiver helps me recover a body, it hasn't yet. The main reason I wear a transceiver during a rescue is so my fellow rescuers can find me if I get buried, and so I can find them.

Dog Searches

The opposite of a high-tech avalanche transceiver is a low-tech dog. Their noses are a thousand times more sensitive than ours. Of the nine avalanche fatalities I responded to, three were found with dogs, two with beacons, and two with probes, and two were at least partially visible on the surface.

Kids at Play

It's Saturday, and I'm working in the first aid room at Brighton Ski Resort. My shift has me working in the aid room from noon to one. We have a patient with a fractured arm, another with a sprained ankle, another with an injured knee, and a sixteen-year-old girl with altitude sickness—in short, a typical Saturday afternoon.

Over the radio, I hear someone report an "out-of-area avalanche with two possible down." Brighton's protocol is that avalanches be reported on the radio as a "10-33," or a

"10-33-B1" if there is one confirmed burial. I think the code is used so that if Joe Public overhears a radio report, he won't freak out over the word "avalanche." In reality, avalanches that affect the public inside ski areas are *extremely* rare—a customer has never been buried at Brighton. In any case, the patroller who calls it in is calm, as is the patrol director, Patrick Eibs, who asks a few questions and gives initial instructions.

I continue to care for the patients in the aid room for about five more minutes until I hear somebody on the radio say, "Has anybody called Lane?" referring to Sgt. Lane Larkin.

I tell the lead person in the aid room that I'm heading up the mountain to learn more about the avalanche. I flash the bright yellow sheriff's insignia embroidered on my fleece vest. It's purely coincidence that I'm wearing my sheriff's office fleece, because when I'm patrolling, I'm working for Brighton, not the sheriff. But it's a warm vest—I wore it inside out to hide the logo.

While riding up the Crest chairlift, I hear Patrick say he is just leaving the top shack. Patroller Rob Nicolich reports that like me, he's a few minutes from the top of Crest.

When I get off the lift, I see a half-dozen patrollers assembled. I grab my backpack from the top shack, which is sort of a log cabin affair. A radio call requests the portable ski toboggan. Like the litters used by Search and Rescue, the portable toboggan separates into two halves that are mounted on a backpack frame. It also has two long handles that attach to one end of the assembled toboggan, so that a rescuer can stand between the handles and ski the patient down the mountain.

A fellow patroller helps strap my shovel, avalanche probes, snowboard, and backpack to the toboggan; then the whole heavy, awkward shebang goes on my back. I'm also wearing a ski patrol fanny pack full of medical supplies, so the whole effect is something like a gypsy wagon moving across the snow.

I walk over to Rob, who is also gearing up, remind him that I'm also on Search and Rescue, and ask if I can go in. I again flash my embroidered insignia. It's a slightly awkward moment, because as a weekend patroller I'm not authorized to go on out-of-area rescues. As a Search and Rescue member, I am, but I need permission to shift hats from patroller to sheriff's employee.

Rob calls Patrick on the radio and says, "Steve Ar-kill-ees"—nobody can pronounce my last name—"the 'National' who is on Search and Rescue, is equipped and ready to come in. Is that okay?"

Patrick replies, "Yeah, send him in."

Rob referred to me as a National, because I'm a member of the National Ski Patrol. It's not always clear whether the paid patrollers view us unpaid "vollies" as peers or as inferiors. In this case, it's clear that Rob and Patrick are all for my going in.

With Rob is a young man, Doug, who was with a group of friends when the avalanche struck; he snowboarded out to get help. Rob asked him to come with us since he saw it happen and may be able to give us helpful information.

As we start hiking from Crest Top, we pass two signs. One reminds us that we're leaving the ski resort and there are no avalanche control or ski patrol services beyond the sign. The other, a whiteboard, reports that today's avalanche danger is "considerable." It isn't a good day to be in the backcountry.

Each morning, based on the Utah Avalanche Center's "advisory," we update the whiteboard. We also set up a rope fence to clearly identify the ski area boundary. This fence has a narrow opening, so that anybody leaving the ski area is forced to pass within a few feet of the signs. I'm wondering whether today's victims read the signs.

We pass through the gate and begin the slog up Pioneer Ridge. The ridge begins at 10,000 feet and climbs several hundred vertical feet. It's steep, but it feels steeper and longer with my big, unwieldy load. Still, the snow is firm, and we can boot up the ridge without punching through and sinking.

I'm sweating heavily as we work our way up the ridge. The thin air and heavy load have upped my heart rate, and before long I realize that I am under-hydrated. I've learned that being tired, loaded down, and under-hydrated is not a big deal in the short term, but over time, you begin to pay. *Suck it up, Stevie,* I tell myself. *You'll be on-site within an hour.*

After climbing for about twenty minutes to a small summit, I get to click into my snowboard and ride a few hundred feet down the ridge. Today I chose to patrol on a board—not nearly as versatile as my telemark skis, but then, I didn't expect to be heading into the backcountry.

The top-heavy load on my back makes the narrow ridge awkward. The pack frame doesn't have a waist belt or a chest strap, so the whole load shifts as I bounce over the uneven terrain. I stop and Rob ties a cravat—a large triangular bandage—around the bottom of the frame to act as a waist belt. Life has just gotten easier.

We then drop down a steep, thirty-foot-wide chute and begin our traverse onto avalanche terrain. As I ride out of the bottom of the chute and onto the narrow traverse, my snowboard slips below the just-cut track, and I sink into the snow above a tree. I can't move back or forward. My ski poles, which I brought to make traveling on a board in the backcountry easier, plunge to their grips in the soft snow. Taking off my board, I sink to my waist, still with all the junk on my back. After an exhausting several minutes of squirming like a worm, I manage to get around the tree, get my board back on, and continue the traverse. Dripping with sweat, I hear that patrollers and avalanche dogs are already on scene.

This is classic avalanche country. I am crossing a broad cirque thirty-five to forty degrees steep. It's a prime avalanche slope—in fact, it has the same exposure as the slope that just slid.

Standard backcountry practice is to traverse any questionable slope one at a time, with your partners acting as spotters—and, ultimately, rescuers—if the slope should go. But Doug, who has just completed the traverse, seems to have forgotten my request to look back and spot me. Rob, meanwhile, has descended to a lower slope to help an avalanche dog that's stuck below us in deep powder, so I'm on my own. When I wake up tomorrow, I'll shudder a little at how exposed I was to a secondary avalanche.

As I begin the traverse, a man skis below me. He is alone with a small backpack and is dressed in typical ski clothing. I call down to him, "Are you part of the rescue?" He says yes and proceeds across the traverse. In hindsight, I should have asked if he had an avalanche transceiver, and made sure we took turns spotting each other as we crossed the terrain. I wonder if he read the "considerable" sign as he left the ski area. Weeks will pass before I learn that Patrick, the patrol director, knew that this slope had recently slid, and had briefed the others that the hazard to rescuers was acceptable.

After traversing the cirque, I arrive at the slide. A big one, it began a few hundred vertical feet above me and ripped down the mountain for about a thousand feet. I can see that the crown, which is the headwall where the avalanche fractured and cleaved away, ranges between three and eight feet high; the debris field is almost three hundred feet wide.

Later we will learn that twelve teenage snowboarders rode the Crest chairlift and left the ski area by hiking along the same route we used. They, too, passed the unmistakably clear avalanche warning signs, hiked the ridge, and traversed to the Dog Lake area, where they planned to build a jump.

Two of the snowboarders told their friends that they wanted to hike up higher so it would be easier to get to the jump site. After hiking, they strapped on their boards and began traversing a very steep (50-degree) slope. The snow fractured above them and took them for the ride of their lives. Their friends watched them disappear in the white cloud.

None of the young snowboarders wore an avalanche transceiver. They did have a half-dozen shovels, which they had planned on using to build a jump, not dig up buried friends. One of the witnesses called 911 while Doug snowboarded out for help. Everyone else started frantically digging random holes in the snow in a futile attempt to find their friends.

When I arrive, I see patrollers from several ski resorts scattered across the slide. Although we responded as quickly as possible, it was more than an hour before the first rescuers arrived.

I can see dogs from Alta, Brighton, and Snowbird working the avalanche debris. Their handlers give commands and watch like hawks for any indication that their charges smell a buried person. Patrick is working both his big German shepherds, the "mother and son" team of Brandy and Jake. I ask him where he wants me, and he tells me to work with Brandy while he stays with Jake.

Several hundred feet below me, a line of fifteen people is searching with avalanche probes— ten- or twelve-foot poles that look something like a tent pole. The half-dozen sections of aluminum tubing are strung together with an internal cable—a design that allows the probe to be carried as a small package and assembled in seconds. The rescuer holds an end section and pulls on the cable, and with a flick of his arm, the tubes connect into a long, rigid rod. The search line is a motley collection of teenage snowboarders standing shoulder to shoulder, each holding a probe. They systematically plunge their probes deep into the snow and then extract them, and at a patroller's call of "Step forward," the line advances. Again the probes punch into the snow and are withdrawn, and the process repeats. Looking at the line, I'm confident that if there is a body under them, they'll find it. As the line advances, patrollers mark the sidelines with small wands tied with bright orange flags. This will help us track the areas that have been searched.

My probe is strapped to my collapsible shovel, which is part of my bulky cargo. I remove and assemble the probe with a flick, and again hoist the toboggan, snowboard, and pack onto my back, keeping one eye on Brandy as I work.

At Patrick's request, fellow patroller and Search and Rescue member Andy Peterson has gone up the slide path to act as a spotter and get an overview of the operation. If another slide comes down—which looks pretty unlikely, since everything that can come down already has—Andy will alert us over the radio.

Rob, a few hundred feet below me, calls up: "Steve, they need the toboggan down below."

"Where?" I ask.

"Down there where Brandon's doing CPR."

I don't see anyone doing CPR, but with the toboggan on my back, I plunge-step down the chunky debris, ski poles in one hand and probe in the other.

As I near a group of people, I see two patrollers next to a small tree. They're working on a kid about eighteen years old. It looks as though the avalanche wrapped him around a tree. One of the patrollers, George, is performing rescue breathing with a pocket mask. The other, Brandon, is doing chest compressions. Brandon counts his compressions, "One-and-two-and-three-and-four-and-five," and then pauses while George gives a breath through the pocket mask. I watch the patient's chest rise and fall with each breath.

I ask if either of them wants a break, and they say no. A bystander—I think it's the guy who skied past me on the traverse—helps me set up the toboggan by joining the two halves and attaching the handles.

Before moving the patient to the toboggan, I have to dig out his left leg, which is buried vertically. I dig down two feet to get to his foot, which is rotated unnaturally. I grab the knee and pull, but the foot doesn't move. Digging deeper to free the leg and foot on all sides, I again grasp the obviously fractured leg below the knee and pull it out of the snow. I align the now extracted leg with its mate. Grabbing the kid's Gore-Tex pants near his hips, I wait for Brandon's count of three; then we heave the body onto the toboggan, where Brandon and George continue CPR.

I again ask if either wants a break. They are both sweating. This time Brandon says yes, and I take over chest compressions.

Although I want to believe I'm making a difference, I don't think we're going to bring him back. Still, even though he's been down for two hours, protocols say that once we start CPR, we must continue.

There's a nasty odor. It isn't the smell of feces—more sour, like puke. It must be air getting jostled up from the victim's stomach. I will smell this in flashbacks for days.

George points out that we are getting good chest rise and that the patient's skin is staying pink. Optimism is not a bad thing. I'd like to think we are keeping him perfused with oxygen, but I'm guessing this is a corpse we're working on.

Between breaths, George says, "I wish they'd just get the helicopter here." He says it a couple of times. The victim's forehead and the side of his head are bluish-scarlet. His femur is obviously fractured. He hasn't had a heartbeat for at least an hour. We've done what we can.

When the helicopter lands near the base of the slide, we interrupt CPR so Brandon can quickly take the patient down in the toboggan. I ask about continuing CPR during transport, but he says no. I know why; he knows why. They'll "call" him as deceased at the helicopter. With that formality out of the way, a patroller will ski the body to the base of the mountain.

I find my probe, which has been passed around among searchers, and am walking toward a cluster of trees when a patroller seventy feet uphill from me starts yelling, "Shoveler!" A dog near him is barking and digging in the snow. I run uphill about half the distance to him and then throw him my shovel. Patrollers from all directions are also running across this uneven field of jumbled snow. Within thirty seconds, a dozen people are digging frantically.

Before I see the body, I smell that same odor again. The victim is buried under two or three feet of snow, on his left side. I climb into the hole and help roll him over. As he flops over, I see that blood has been running from his nose. I check him for signs of life. Turning, I see Doug, the snowboarder who watched his friends get buried, choke up. I walk over to him and take him away from the body. He crouches down and begins to sob quietly. I put my hand on his back and don't say anything. We stay like this for a few minutes until Gary tells Doug to go join the rest of his friends. It was a good decision to get them all together, away from the body. Looking at the lifeless body isn't doing him any good, and being with his friends, who are still among the living, will begin the long healing process. It's good for me, too, since seeing him suffer makes the death more painful for me, and I need to stay a little aloof from the suffering so I can do my job.

We stand around the second body for ten minutes waiting for a nurse from the helicopter to come up. She formally calls him—"Eyes fixed and dilated"—and says the time. We put him in a body bag, first wrapping him in the clear sheet plastic, then zipping him into the light nylon bag, and finally zipping him in the heavier Cordura bag. The medical examiner's report will tell us they both died from asphyxiation.

Rescuers drag the bag downhill through the snow to a waiting snowcat. I climb into the shotgun seat with Brandy, silently scratching her head as we ride back to the ski resort.

We hold our debriefing—fifteen or more of us, standing around two body bags near the edge of a parking lot. The medical examiner goes through the victims' pockets as we talk. A few team members hold blankets to shield the view from the media. My head is swirling with thoughts, but I don't say anything during the debriefing. In hindsight, I wish I had pointed out how well the multiple rescue teams—Alta,

Brighton, Snowbird, Solitude, Search and Rescue, and the public—worked together. Though that isn't always the case in multiagency rescues, it was today.

Back at home, I realize just how exhausted I am. I didn't get to sleep until two last night because of another rescue. I didn't eat lunch and drank only a liter of water in a day spent lugging gear, doing CPR, and packaging dead kids. After the rescue I was pumped with adrenaline; now I'm pooped. I climb into bed and turn on the 10 o'clock news. There I am, loading a body into a helicopter.

A few days later, I realize how intense my tunnel vision was. I couldn't pick the two avalanche victims out of a lineup—not even the young man I did CPR on for twenty minutes. When I rolled over the second victim, I don't think I really saw his face at all. I looked at him, saw that blood had run out of his nose, checked to see if snow was in his mouth and throat, and noted that his eyes were partly open, but I didn't really *see* him. It's a little strange.

Steve and George fight to revive the young man
(Visit MountainResponder.com/Photos and enter Story Code BSA)

Gobbler's Knob

The call comes a little after six p.m. I'm at home, cooking stir-fried veggies and mashed potatoes.

```
This is a callout for S&R on an
avalanche in Mill Creek Canyon..meet
at Nob Canyon in Mill Creek..respond
on Code One Frequency..case#17920.eom
18:18 02/15/03
```

I change clothes fast and explain to my girls and my nephew, who's staying with us on a ski vacation, that this callout probably won't take long. After all, this is the third avalanche callout in a week, and the other two were canceled shortly after being paged.

The sun is setting as I drive toward the staging area in Mill Creek Canyon. From the radio traffic, it sounds as though the command post is being moved to Mill B in Big Cottonwood Canyon, so I turn around and head there. A few minutes later, a follow-up page confirms the new staging area.

While listening to the radio, I hear that Fire is already at the Mill Creek staging area and Life Flight's helicopter has already spotted the party. How in the world did Fire drive the half hour to the Mill Creek staging area, and Life Flight do a flyby, before we were paged? Initially, the 911 dispatcher notified the fire department rather than the sheriff's office. The responding fire captain knew that the call was for an avalanche, yet he rode the fire engine all the way up Mill Creek and dispatched a helicopter before notifying the sheriff's office. No skis or backcountry skiers, no avalanche beacons, and no avalanche rescuers. What was he thinking—that they were going to walk up the snow-covered mountain in their bulky fire-resistant turnout gear? I don't have the skills or experience to pull somebody out of a burning building, and most firefighters haven't logged the years of experience needed to travel safely in high-risk avalanche terrain.

My drive to the staging area is stressful. Traffic is moving slowly up the canyon. I wish I could turn on overhead lights and get past the cars, but the sheriff's office doesn't permit it. *Damn,* but it's hard to be patient when somebody is dying—even when patience is the prudent thing.

I hear Sergeant Larkin on the radio, calling for a second helicopter and ski patrollers with avalanche dogs. Those are the appropriate resources for this call. The sun has already set on this short January day, and darkness is going to complicate things.

The Life Flight pilot reports that he can see several people but that he can't land on the steep, snow-covered mountainside. About five minutes later, he calls to say that he's found a location where he can land, but it's several hundred feet below the stranded party. I'm thinking, *Mister, if you can get us within a thousand feet of the victims, you'll still save us hours of hiking on skis.*

When I arrive at the staging area, the helicopter has already landed at the big parking lot near the Mill B trailhead. It's dark except for a rising moon and the flashing lights on the swarm of emergency vehicles. I see Commander Gary Banks's truck on the side of the road.

As I'm parking, I see another rescuer suiting up in hiking boots and snowshoes. I pull on my ski boots and grab my poles and already-skinned telemark skis. After jamming a pile of extra clothing into my backpack, I hurry over to the other side of the street to hear an update.

The initial report is that there is one confirmed fatality and four buried victims. *Four buried?* I'm stunned. Backcountry winter travelers are taught that only one person at a time should be exposed to avalanche risk, so that the remaining people in the party can mount a rescue. Clearly, mistakes happen, and we occasionally have two avalanche victims, but one confirmed fatality and four people still missing and presumed buried? This is monstrous.

I run back to my truck for a trauma kit, a second oxygen bottle, additional airway gear, and the football-shaped bag valve mask, or BVM, that will allow us to breathe for a patient—just one patient. As I run, I know I should walk and remain calm—staying calm and relaxed almost always ends up being faster than running, but the scope of this mission takes over my thoughts. A minute or two either way may actually make the difference between life and death.

I walk toward Gary's truck to see what he needs. As I do, I hear him ask, "Is Steve here?" I am a little embarrassed and also flattered—the boss asking for *me*? I walk up and Gary asks if I am ready to go. I am; we will be the first team flown in.

Returning to my truck, I take out a roll of white medical tape and secure my ski poles to my skis, to minimize the loose gear in the helicopter. Then I grab the bundled skis and poles and my stuffed backpack and schlep everything back to the helicopter. Someone from the flight crew loads my gear.

A member of the flight crew—I think it's the paramedic—shows me where I'll be sitting. He explains that the landing zone will be very small and that when we get out, we should kneel next to the ship and not move. He will unload our equipment, and we are to remain crouched near the ship until they fly away. This is standard skier-helicopter operating procedure, but these folks work with lots of different rescuers and frequently repeat the instructions. That's okay with me— I'm glad he's familiar with inserting skiers.

I'm feeling the buzz of going into a potentially dangerous situation—certainly an unknown one—in an effort to save lives. The excitement is different from the excitement of a roller coaster ride or a great ski run. This "excitement with a purpose" includes the adrenaline high, but it also involves an intensity of focus. It isn't excitement with a grin—this is excitement with a narrowed mouth and focused eyes.

Climbing into the ship, I catch the familiar whiff of aviation fuel. I put on my helmet and look into the darkness, and as the ship takes off, I think about the risks of this mission. We're flying in a helicopter, in the dark, in the mountains, into terrain that just avalanched, to look for up to four buried skiers. Staring out the window into the night, I'm aware of the occasional patch of distant city lights rolling by beneath us. I assume we are circling to gain altitude, but I've lost all sense of direction—I just hope the pilot hasn't.

I rarely think about my own mortality during a mission, but I do tonight. Sure, there is a chance, albeit a small one, that my daughters might lose their father tonight. Of course, the same could be said about every dad who is driving home from work on the freeway, or even the ones who take the train. I have an unspoken promise with myself that my kids should not lose their dad while he's out trying to save somebody else's. The thought passes quickly. There is a good chance that we can help people tonight, and a minuscule chance that I will be killed. Without rescuers, our victims are SOL, and for whatever reasons, we've assembled a set of skills that just might make a difference. As my meditation on death passes, in its place is a certain knowingness that this is what I'm supposed to be doing. A calm born of accepting one's destiny sweeps over me. I have this fatalistic sense of "whatever happens tonight is what's supposed to happen," because I am supposed to be right here, now, doing this.

The chopper is loud, and I forgot my earplugs. I press my index fingers over my ear canals, trying to block out the sound. I'm sure I look goofy as hell, but ear damage is a certainty, whereas looking dorky is merely a passing circumstance.

As the chopper wends its way into the sky, I see an almost full moon looking back at me. The snow below us is lit up by the copter's massive belly lights. The *whop-whop* of the rotor blades cutting into the cold air penetrates my body as we are pulled into the sky. Occasionally, the bumping ride coincides with a beeping sound that is apparently warning the pilot of something or other. In my fatalistic mood, I ignore it.

I'm focusing on the mission. Will there be victims on the surface of the snow? Will we locate buried victims using our avalanche transceivers? Using our probes? How deep will the victims be buried, and will the snow come off easily or like concrete? If we uncover a victim who isn't breathing, do we stop to attempt resuscitation or keep searching for more victims? I can't quite wrap my mind around two people searching for four avalanche victims. We'll just do what we can.

After about ten minutes, the pilot works his gangly bird into a gentle landing on the snow. The last few inches of the landing take at least a minute—the pilot is taking great care to see that we don't sink into, or slide off, the snow on this knoll-like LZ.

The paramedic climbs out the left-front door, goes around the front of the ship, and opens our door. Gary and I climb out and scrunch down next to the ship, securing our skis, poles, and backpacks as they are passed to us. The pilot looks at us over his right shoulder as he lifts off. The prop wash sand-blasts us with snow as the ship departs. We are deep in the mountains, alone with just the moon.

Ironically, I have no clear idea where we are—I'm not even sure which canyon we're in. We were first told to stage in Mill Creek Canyon, but we were rerouted to Big Cotton-wood. I heard people saying we were going into Cardiff Fork, but knowing that the mission was safely in Gary's hands, I was busy getting my gear together and never heard our desti-nation. When I ask Gary where we are, he gives me a big grin, chuckles, and tells me, "Steve, my man, we are in Mill Creek, below Gobbler's Knob." I smile back. What a goof for not knowing where I am!

Looking up the mountain, we can see a light in the dis-tance. It must be members of the group. With the skins already on my skis, I attach the skis to my boots and strap on my backpack as Gary pastes on his climbing skins. He calls down to the command post telling them we can see a light and that we are about thirty minutes away—it looks more like an hour to me.

The climbing skins on the bottoms of our skis are like a sort of bristly "snow Velcro," allowing our skis to slide forward as we "walk" uphill, but not letting them slide backward. Attached to the tip and tail of each ski, the skins stick to the bottoms of our skis with an always-tacky glue. Our ski boots attach only in front, to hingelike bindings so our heels can lift as we trudge up the hill.

As we begin skinning, the brush and willow branches that protrude through the thin snowpack grow tighter and thicker, making travel harder. The surface of the snow is a breakable crust that doesn't quite support our weight, so we drop several inches through the crust with each step. We discuss moving onto the avalanche debris below us. I think we should, because it looks as though our current route will soon be too steep and brushy, but I voice my opinion tentatively. Gary prefers to hold on to every foot of altitude we've gained.

After ten minutes, he yells ahead to me, "Hey, Steve. Call up the party and have one of them come down and meet us." He's trying to save a few precious minutes. I call back to him that I will in a few minutes. The truth is, I think I'm too far away to communicate with them. A few minutes later, I realize I have arrived at the group.

Gary, the Experienced One, was right on both counts: staying high and going through the brush was the best route, and I could have called out to the party sooner. I hope I gain his decision-making ability through experience, though my plastic left eye means I'll never have his visual depth perception.

"Hi. I'm Steve with the Salt Lake County Sheriff's Search and Rescue Team. What happened?" I ask.

Their solemn reply: "Our friend got caught in an avalanche and died."

I ask if anyone else is buried. They shake their heads and tell me no. I restate the facts for clarification, "One person is dead and nobody else is hurt, buried, or missing, is that correct?" They're obviously sad and not in a talkative mood. They nod their heads. *Whew!* Big relief! I came in expecting to do a rapid search for up to four buried victims, in a challenging medical situation where the two of us would be unable to provide medical care for multiple patients. And now I learn there is "only" one fatality and no significant medical care is required.

To my right, a pit has been dug in the snow. It's about eight feet long and four feet deep, and there's a skier in the bottom. I climb down into this gravelike excavation, pull back the hood on the skier's parka, and look at him. I reach up and check for a carotid pulse on his neck. The tips of three of my fingers rest against his neck. At first I think I feel a pulse. I've had it happen before, reaching a patient after a strenuous hike or ski, and at least initially mistaking the blood pounding through my body for the victim's pulse. After all, this guy is cold and stiff, and his friends told me he has been down since 5:30—well over two hours. I check his pulse for a full sixty seconds just to make absolutely certain he isn't alive and just hypothermic with a very slow pulse. I open one of his eyes and shine my headlamp into it. It's dilated and does not react to the light. I repeat this on his other eye.

I climb out of the hole just as Gary is arriving. I tell him the victim is pulseless with fixed and dilated pupils and that he is the only victim. Gary calls down to the command post to inform them that there is only one "Echo" and no burials.

We will later learn that the victim triggered the avalanche at an elevation of about ten thousand feet on a steep north-northwest–facing slope. The initial avalanche was only a foot deep and a hundred feet wide, but it carried him fifteen hundred vertical feet. As it traveled, it released secondary avalanches on either side until it grew to more than three hundred feet wide. This large volume of snow then funneled into a narrow gully, burying the skier under four feet of dense snow. His partners did all the right things: located him using their avalanche beacons and attempted to revive him using CPR.

The victim, forty-eight-year-old Allan Davis, was a volunteer observer for the Utah Avalanche Center. He and his friends were experienced backcountry travelers. Unfortunately, Allan got into a localized area with higher avalanche hazard. This illustrates the complexity of avalanche forecasting, even for experienced backcountry travelers, because the hazard can vary tremendously within a few hundred feet.

As I am talking to his friends—two men and two women—Life Flight drops off two more rescuers: Andy Peterson and Rob Hunter. They bring in a backpack-mounted toboggan like the one used by Brighton.

We are still standing on the avalanche debris next to the pit containing the dead skier. Normally, avalanche debris is a safe place to stand, because the snow has already avalanched and is unlikely to do so again. But Gary knows that there are several slide areas that join together above us, which could potentially come down this path. One of Allan's friends tells us that the entire upper area already slid, but just to be on the safe side, Gary has me kick steps up a sidehill and move the four friends out of any potential avalanche path.

We have a tough time lifting the victim up out of the snow. It always amazes me how a body sticks to the snow, even when unburied. Even a victim who is on top of the snow often still can't be lifted. In this case, though, it also turns out that one of his legs is connected to a buried ski. We dig until we reach his ski and release the binding. The ski will be left behind.

While we are releasing his foot, Gary does a quick trauma exam, looking for obvious fractures. It makes sense that we do trauma exams on seriously injured patients, but we often do them postmortem, too—it helps us understand what happened and improves our assessment skills. I wonder what the victim's friends think as they watch us rechecking him for injuries.

The four friends are out of food and water, and we give them four of our six water bottles. They are appreciative and repeatedly thank us for coming. I offer them several granola bars, which they politely decline. I finally convince them that the food was purchased with their county tax dollars and that they really should take it. It is amazing how some people are so grateful for even the smallest assistance, while others act as if they were entitled to our services. These folks definitely fall into the former group.

The helicopter lands again, this time bringing rescuers Laurie Jess and Andi Wasser. Rob Hunter skis the four friends down to Laurie and Andi, who will ski with them to waiting snowmobiles. He then returns to help transport the body.

We load the body into the toboggan. He is tall and barely fits. I climb between the toboggan handles—the horns—as Andy takes the tail rope. Taking a half-wrap of the rope behind his back, he holds it tight against his hip to provide a safety belay, letting me advance a mere fifteen feet, at which point I stop and hold the toboggan secure while he hobbles down to me, and then I progress another fifteen feet. We are limited to these short pitches because of the short tail rope and the steep avalanche debris. The debris is frozen hard and covered with head-sized chunks of ice, so we can't really ski it. Instead, we sideslip a few feet and then step over the chunks of ice. We repeat this roped technique until we're off the steeper debris.

We are working down hellish terrain: avalanche debris, gullies narrower than the length of our skis, breakable crust over hollows, branches, steeps, and sidehills where the toboggan wants to yaw sideways or even roll. The size of our passenger adds to the challenge. Andy and I patrol together and get frequent experience with toboggans, but on relatively tame terrain at a ski area. Tonight's toboggan handling is a daunting task. Our progress is slow and safe, but no one would call it pretty.

At times the sidehills are so steep that we have to connect additional ropes to the uphill side of the toboggan. Then Andy, Gary, and Rob go uphill, holding the ropes in their hands or snubbing them around trees, to help maintain control of the toboggan. Letting the toboggan take off downhill with its uncomplaining passenger would be good slapstick in a movie, but it's our nightmare scenario and constantly lurking as a possibility.

We come to a steep twenty-foot bank that we have to go up, and the four of us can't pull the loaded toboggan uphill through the soft snow. We don't have pulleys or a long rope. So, à la MacGyver, we tie several short pieces of rope and webbing together to make a longer line and use carabiners as pulleys to create a three-to-one mechanical advantage system. With two rescuers pulling with all their strength on the rope, and two rescuers keeping the toboggan from tipping over, we advance it, one grudging foot at a time. This process is exhausting—it takes us forty-five minutes just to get up this stinking hill.

There are times when I think we may need to leave the body until tomorrow. Gary doesn't want to—he's concerned that although the avalanche danger is "only" rated moderate, which translates as "Avalanches are infrequent but possible," this accident shows that the hazard is still pockety. If we come back in the morning, we'll risk having other people ski above us and potentially start an avalanche that buries us. He thinks it's safer to bring the body out tonight, when we don't have to worry about anyone above us.

After almost four hours of dragging the toboggan, we meet rescuers who hiked in on snowshoes from below. They give us their water and energy bars, which are very welcome—we've been out of water for hours. With fresh legs, they take over the toboggan detail and are quickly out of sight.

Andy, Rob, Gary, and I ski down a small ridge to a road that has been closed all winter. Several snowmobiles and the team's snowcat are there. We're offered a ride in the cat, but Andy and I decide to ski down the road. The gas-powered machines and their noise leave with the other rescuers, and Andy and I ski down the road. There is no sound but the hiss and scrape of skis over moonlit snow and ice. Without speaking, we share the pleasant feeling of accomplishment through hard work, and have no doubt that our avalanche victim would be pleased that his passing should grant us this lovely midnight ski.

When we get to the main road in Mill Creek, we find an amazing array of emergency vehicles, firefighters, and rescuers. We also see crying family members. The rotating red and blue lights dancing across the snow give the scene a dreamlike feel. A few people come up and thank or acknowledge us. I'm feeling good. It's a beautiful night, but for the death and grief.

Since we skied into Mill Creek Canyon and our vehicles are back in Big Cottonwood where we boarded the helicopter, a deputy shuttles us between the canyons. I get home at 2:30 a.m. The callout began eight hours ago, when my pager sounded. So much for my confident pronouncement about being home soon. The mashed potatoes are still on the counter.

Ignorance Kills

On December 11, 2004, the sheriff's office got a call from a concerned family member saying that two hikers were overdue. Deputies took a description of their car and found it parked near the entrance to Mineral Fork, a local canyon. The Search and Rescue Team was paged, and arrived at the trailhead at about 9 p.m.

Existing Hazard

The call for overdue hikers—snowshoers, actually—comes with a mood of foreboding. Two people have already died in the Utah backcountry in the past twenty-four hours. Last night our team helped bring out the body of Zachary Eastman, a twenty-three-year-old backcountry skier, killed in an avalanche between the Brighton and Solitude ski areas. This morning a snowmobiler died in an avalanche in a nearby county. Three other snowmobilers were caught by avalanches today and survived.

I'm still baffled at how people end up in this situation. Sure, maybe they didn't hear about today's accidents, but last night's fatality was all over the TV news. And today's avalanche report could not have been more explicit in broadcasting the hazard:

> *Especially avoid steep north through east facing slopes, even at mid and lower elevations where the weak faceted snow was weakest before it got buried. I wish I could continue to call it a HIGH danger today, but it doesn't quite fit the definition of HIGH, so I won't call it HIGH. It is just barely on the CONSIDERABLE side of HIGH. Instead, it's a very HIGH-pucker-factor CONSIDERABLE. Get the point? This means that human triggered avalanches are probable and natural avalanches are still possible. The bottom line today is that no way, no how should anyone jump into steep terrain. Just give it a rest this weekend.*

I'm running this rescue. From the trailhead I call Bruce Tremper, who is also the director of Utah Avalanche Forecast Center, on a satellite phone—cell phones won't work here. I explain our situation—two overdue snowshoers in Mineral Fork—and ask him if he thinks it's safe for rescuers to go in. Bruce's reply is clear: "I sure wouldn't go in." He will later describe today's conditions as "unstable as a brick on a stack of potato chips."

Although I don't want to ignore his expert advice, rescuers James Taylor and Keith Sauter assure me they can go in safely on a small morainelike feature in the center of the narrow valley. I agree to their doing a quick inspection of the area to see if they can pick up a signal from an avalanche transceiver and to make sure there isn't an injured survivor on the surface. Finding a protruding hand—or, worse, a head—the next morning would be a nightmare. I don't want to find a corpse tomorrow that could have been a save tonight. I also don't want to get a rescuer killed.

James and Keith don their skis and follow fresh snowshoe tracks up the narrow canyon. When the canyon opens up, the tracks disappear into recent avalanche debris. They search the perimeter of the debris and can't locate any tracks leaving the area. They sweep the slide with their transceivers but can't raise a signal. Using their headlamps, they search for visual clues on the surface. Nothing.

We fly an avalanche expert over the area, who agrees that conditions are too dangerous to bring in additional searchers tonight. The slide fell more than a thousand vertical feet, covering twice that distance on the surface. It is more than four hundred feet wide.

The next morning, snow safety experts from Wasatch Powderbird Guides fly the area with their helicopter. To stabilize it for the incoming rescuers, they throw forty-five explosive charges out the helicopter door into the "hangfire" snow, lying above the existing avalanche, which hasn't yet slid. The snow from some of these new avalanches spills onto the previous debris.

Search and Recovery

Once the hazard has been cleared, several teams of rescuers are airlifted to the accident site. We search the area using avalanche transceivers, avalanche dogs, and probe lines. We also use an electronic device called a RECCO detector, which can detect special reflectors that are attached or sewn into clothing or equipment. The detector can sometimes detect other electronic devices, such as a cell phone or GPS unit, at shorter ranges.

At 11:30 a.m., a searcher on a probe line hits something. After forty-five minutes of shoveling, I uncover thirty-seven-year-old Melvin Dennis, buried under four feet of snow.

We spend the rest of the day searching for Bruce Quint, without luck. Before leaving, we mark the avalanche perimeter with orange bamboo poles so we will know the location of the original avalanche if it snows tonight.

Waking up the following morning, day three, I think about the Utah County avalanche a few years ago, in which it took more than three months to locate one of the victims. I'm hoping we don't end up with a similar situation.

More than fifty people have turned out for today's search. Besides the helicopter pilots, Search and Rescue Team members, and a dozen ski patrollers from Wasatch Backcountry Rescue and their search dogs, we have reporters and cameramen from a half-dozen media outlets, the Red Cross serving food, and cops and deputies.

Shortly after sunrise, we fly in several dogs and their handlers, hoping that the calm air will help the dogs do their work. Within an hour, one of the dogs, a shepherd named Midas, takes an interest in a nondescript small area in the middle of the debris field, and his handler probes the area. With only a few feet of his twelve-foot probe protruding from the snow, the searcher calls me over. Pressing down on the probe, I feel the slight springiness that may well indicate a buried body. Several other searchers come over to "feel" the find—and the shoveling beings. More than ten shovelers spend an hour uncovering fifty-nine-year-old Bruce Quint. He is buried under nine and a half feet of snow.

Diagnosis

Neither Melvin nor Bruce was wearing a transceiver. Not that it would have made much difference, since they were travelling next to each other and buried at the same time. What *could* have made a difference was a phone call to the Avalanche Forecast Center.

Four avalanche deaths in thirty hours. What a waste.

Avalanche victim buried under nine feet of snow
(Visit MountainResponder.com/Photos and enter Story Code MFA.)

Six

Lost and Found

Most search and rescue teams do a heck of a lot more searching than rescuing. In Salt Lake County, the abundant recreational terrain next to a large population center flip these proportions around—the team generally spends very little time searching and a lot of time rescuing. And occasionally, we get called out when no one's even lost or hurt.

Evidence Searches

A few times each year, in a departure from business as usual, we were called out to search for criminal evidence. Sometimes the search amounted to wandering through fields looking for a purse or personal effects following a murder, but the typical call was to search the Jordan River for a gun. We were rarely successful.

On March 29, 2004, Salt Lake County Sheriff's Sgt. Thad Moore drives me to a bridge crossing the Jordan River. Moore explains that a nineteen-year-old man has confessed to shooting dead an eighteen-year-old. Apparently, the killer calmly shot the victim in the head at point-blank range and threw the gun in the river.

As we're sitting in the sheriff's SUV, a marked car pulls up, and we watch as a plainclothes detective takes a young man out of the backseat. The man is dressed in orange prison clothes, and his hands and legs are shackled with chains. He hobbles along the edge of the bridge and points into the water with his bound hands. The deputy asks him some questions, and he replies, then gestures with his hands as if tossing something into the water. I feel a little chill watching this self-confessed cold-blooded killer calmly explain his actions.

The next day the team is called out to search the slow-moving river. The water is only a few feet deep here, and I had a good (though unsettling) view of the suspect as he pointed to where the murder weapon should be.

Dressed in baggy brown waders and yellow rubber gloves, my teammates and I search the river. Our primary search tools are big, powerful plate magnets, eighteen inches long and eight inches wide, suspended from ropes. As we wade through the river, we hold the ropes and skim the magnets over the bottom of the river, periodically raising them to the surface to check our "catch." In a way, it's a little like deep sea fishing, in that you don't know for sure what you're going to haul up. We come up with railroad spikes, stop signs, hand tools, car keys, cell phones, and a shopping cart to carry it all.

During the search, as we're lowering the magnet back into the water, a single drop splashes up and lands in my mouth. I quickly spit it out and keep spitting for a while, hoping I haven't just contracted God knows what sort of vile disease. Searching an urban river may be more dangerous than working an avalanche or hanging off a cliff.

After several hours of searching, we pull up a rusty shotgun. Unfortunately, we were searching for a handgun, which we never found. Anyway, the guy confessed, and he's behind bars.

Rescuers drag magnets to search for the murder weapon
(Visit MountainResponder.com/Photos and enter Story Code JRS).

Missing People

More often, we were searching for missing people, not guns. There's a scientific method to the process. Initially, we would determine the point where the missing subject was last seen, and send a hasty team to look in the likeliest locations. We would also try to contain the search area by having deputies drive the main roads or having hikers patrol the perimeter. While this was happening, we would use mapping software to divide the terrain that might contain the missing person into search areas. Rescuers familiar with the terrain would then estimate the probability of the missing person being in each area, and the probability of searchers finding a lost person in each area. By multiplying these two probabilities, we could estimate the likelihood of a find and allocate our resources.

Urban Searches

A few times each year, we were called out for a missing person in an urban area. Most of these were resolved quickly. We would find a "missing" child at a friend's house, sleeping behind a couch, or wandering around the neighborhood with friends. Male adults who were supposed to be "out hunting" would sometimes be discovered pursuing a different sort of quarry in a local bar.

Many reports of missing persons involved subjects who were mentally impaired due to a medical condition, such as Alzheimer's disease, or psychologically depressed and suicidal. Sometimes the missing person was the victim of foul play.

Suicide Risk

We had many callouts involving persons at risk for suicide. Unfortunately, most were successful. This suicide call came at five in the morning:

```
Callout for Search and Rescue to
respond on missing suicidal male.
Command Post to LDS Church at 3850 E
Oakview. Cont 821 on Code one
channel. #132614 EOM
5:21 11/01/02
```

When we arrive at the command post, Sgt. Lane Larkin shows us a suicide note, obviously from someone in a psychotic state, that was written last night—Halloween. The author claimed to have slit his wrists, and the note is indeed smeared with blood.

The rescuers are divided into teams of two, to search a well-to-do neighborhood of mountain homes. As we walk through the neighborhood calling subject's name, it feels like a typical urban search—a waste of time. If someone wants to kill himself, I don't think he's going to answer our calls.

While walking up a driveway toward a large home, my teammate Rob Hunter and I see a young man standing by a tree with his back to us. I notice that he's wearing red medical gloves. I'm familiar with tan latex and blue Nitrile gloves, but I haven't seen red ones before.

Rob calls the missing man's name, and the glove-wearer doesn't turn to look at us but instead starts walking away slowly. When we call out again, he takes off running. Rob and I start running after him, just as cops would do. He jumps a chain-link fence, and we're right behind him. In mid pursuit, I make a panting radio call to the command post, giving them a description and direction of travel. I remember, in that moment, reflecting on how I had always assumed that the heavy breathing on those reality cop shows was because the cops were out of shape. Pausing at a small intersection, I look both ways and realize I don't know which way he has gone.

Two-man teams are redeployed starting a few streets away from the intersection where we lost him. As the teams move toward the epicenter, we methodically search the backyards of nearby homes. I'm working my way back to the intersection when I hear sirens and look up to see a fire engine come around the corner and park in front of a big house. Firefighters climb out, looking bored. I walk up to one of them and ask, "Have you heard that the sheriff's office is in the neighborhood searching for a suicidal male?" He grunts and gives me a "No, and who the hell are you?" look as he walks down to the house. I call Command and tell them that Fire is here on a fire alarm. A few minutes later, the firefighter comes running up from the house and starts frantically putting on his turnout gear. Smoke is coming from the back of the house.

Walking down the steps to the backyard, I see smoke coming out of the pool-utility room. On the deck, a trail of dark red blood drops leads into the room.

It turns out that our suicidal male had kicked in the door to the utility room and hidden under an old mattress. During the process, a cardboard box tipped over against a hot tub heater and caught fire, which started the house and the mattress on fire. After the firefighters put out the burning mattress, deputies found our man hiding under it—naked now and holding a steak knife.

It turns out that the unusual red medical gloves I saw were actually the guy's hands, covered in blood from his slit wrists.

And my earlier griping about how we emergency responders should stick to our respective areas of expertise? It cuts both ways. I had no business playing cop and chasing this guy, and it's a good thing—for me, especially—that I didn't catch him.

Seeing God

Quite a few of our callouts were for people with emotional challenges. Some were depressed, some suffered from illnesses such as Alzheimer's or bipolar disorder, and some were downright delusional.

```
This is a callout for S&R on a mental
subject on Mt Olympus Trail Head..
5600 s Wasatch..meet at the parking
lot..respond on Code One Frequency..
case#99248..eom
17:10 08/23/03
```

A fifty-something female, who was out of shape and weighed well over two hundred pounds, had it on good authority that the world was going to end in seven days. After her release from the hospital earlier in the day, God told her to go to the top of Mount Olympus. Her otherwise supportive husband prudently pointed out that neither of them was in good enough shape to make the hike.

But with the end of the world looming, she was not to be swayed. Instead, she convinced her daughter's ex-boyfriend to hike up the mountain with her—he was to carry the ice chest should they become hungry while awaiting the rapture, apocalypse, or whatever was a-brewing. After going a short distance up the trail, the daughter's ex quickly tired of the quest, and called the daughter to say he had had enough. So the daughter went up the trail to take his place in escorting her confused mom to meet God. We became involved when the daughter called 911.

We finally catch up with the duo six hours later at the top of the mountain, 4,400 feet above the valley. In a manic burst of divinely inspired mind over matter, the mother had managed to summit this sizable peak. Happily for all concerned, her manic mood continued and she was able to get back down the mountain on her own power. As we loaded her into the ambulance, I thought, *This poor woman is going to be so sore and tired, she's going to sleep for seven days and miss the end of the world.*

Rapid Response

The Sheriff's Search and Rescue Team was a great rapid-response team for an urban search, because we could have twenty to thirty trained searchers, with radios, our own dispatcher, mapping software, GPS units, and a functioning command post on scene within about thirty minutes. But after the first day, our value to an urban search diminished dramatically because, especially on a big public search, the thousands of searchers who turned out provided many more eyes and ears. Our real value lay more in the speed with which we could respond and in our understanding of search logistics than in the size of our team or the area we could cover.

Foul Play

This looked much like any other missing-person page:

```
S&R needed to assist SLPD looking for
a missing adult female 1025 at the
lower part of Memory Grove come 10-8
on SO freq code one channel eom
07/19/04
```

When we arrived, a detective from the Salt Lake Police Department told us that a woman had gone on her customary morning trail run and hadn't returned. We spent the day scouring the narrow canyon without luck. When the husband came to the search area, shook my hand, and tearfully thanked me for our help, I felt an increased urgency to find his missing wife. The pathos of her disappearance got the public's attention, too, and within mere days more than a thousand volunteers were searching for Lori Hacking.

Thirteen days later, husband Mark Hacking would be arrested on suspicion of aggravated murder. Police would conclude that he shot his pregnant wife in bed and disposed of her body in a dumpster. It would be found two months later in the county landfill. While we searched for his wife, Mark was spotted buying a new mattress to replace his bloody one.

And then there was this call for a youngster who disappeared five days earlier:

```
ADVANCE NOTIFICATION of callout for
07/21/06. S&R units meet at command
post in lower Memory Grove 1600 hrs.
Searching City Creek Cyn area. EOM
22:00 07/20/06
```

Again we were being requested to search the wooded Memory Grove area, the same location where we searched for Lori Hacking. Some five thousand volunteers had been searching the urban areas since five-year-old Destiny Norton's disappearance. She was found by police eight days after disappearing, less than a hundred feet from her home, dead, in the basement of a neighbor's house. The neighbor is spending life in prison without parole.

Misinformation

When a mission begins, the team has no idea how it will turn out. Just as a reported child in a river can turn out to be climbers on a rock wall, reports of a missing person, more often than not, turn out to be the result of poor communication. Other callouts, like Lori's and Destiny's, turn out to be major media events.

I'm cooking eggs for my two girls when these back-to-back pages come in:

```
S&R callout (1 of 2). Assist CityPD
on kidnapping, 1509 E Khristianna Cir
420 N. Command Post in pkng lot of
Shriners Hosp, respon on code-1.
(eom)
07:14 06/05/02

S&R (part 2) Suspect is armed, s&r
units w/certifications, bring your
firearms, CityPD will be providing
armed security for S&R units also. Cs
63405
07:16 06/05/02.
```

I reread my pager and look over at my children. "I'm not going if I need a firearm," I mumble to myself. In years past, team members were required to be handgun certified by the sheriff's office. That policy was reversed before I joined the team, and the sheriff's office no longer allows rescuers to certify through Peace Officer Standards and Training. Of course, in Utah, if you have $65.25, can fog a mirror, and don't have a criminal background, you can get a permit to carry a firearm. A curious twist is that these days, rescuers are not allowed to carry a handgun on callouts, even with a permit. That's probably okay—after all, we're out there to save lives, not take them.

Looking at my pager again, I conclude that I shouldn't die searching for someone else's fourteen-year-old. Ignoring the page, I drive my daughters to school. Then, while parked next to the curb at their school, I reread the pages. "Kidnapping"—somehow, that word carries more weight than the words "Suspect is armed." I couldn't *not* search for a kidnapped fourteen-year-old when my ten-year-old and fourteen-year-old are safely in school. I drive to the staging area.

When I arrive at the command post, the street is already filling with media vehicles. By tomorrow, it will be almost impossible to drive anywhere near the command post, thanks to the swarm of vans and trucks with their satellite dishes perched on telescoping poles.

A fellow rescuer and I are paired with an officer from the Salt Lake City Police Department. The three of us go to the front door of homes in the neighborhood, explain who we are, and ask if they heard or saw anything unusual last night. With permission, we then search around their house and yard, looking in storage sheds, behind bushes, in parked cars, and under pool covers. Occasionally we find normally innocuous stuff, like a fresh Wendy's burger wrapper in the bushes, and radio the information to the incident commander, who will send in detectives.

After a few hours of door-to-door urban searching, I get assigned to a team of rescuers and sent to search the nearby foothills. When we reach the top of the first hill, we look back to see the police officer who was assigned to our team, gassed out about halfway up. It's a case of different strengths: cops are good at confronting bad guys, handling domestic disputes, and other situations that can get pretty hairy and scary; rescuers are good at things like climbing, providing medical care, and cranking up steep hills.

One of the rescuers, Liz Schulte, has her search dog, Moxie, with her. As we comb the well-worn trails in the nearby foothills, Moxie alerts on the frequent signs of human scent; we follow her on various side spurs and then return when the scent peters out.

Throughout this hectic morning, I hear several reports that the kidnapped girl has been sighted. There's a report that she was seen using a pay phone at a convenience store, a caller who claims to have seen a girl in the backseat of a car on the freeway mouthing the words "Help me," and rumor of a two-million-dollar ransom demand. None of these rumors, most of which I overhear in the command post, ever make it into the media.

One caller tells the police that the kidnapped girl is at a construction site, in a box, and still alive. This results in all members of the team being called back to the command post, to shift our search to residential and commercial construction areas. We spend the afternoon searching a new eye center and cancer institute being built near the University of Utah Hospital. After searching every box on the property, I don my medical gloves and climb into the construction dumpsters, praying that I don't find a fourteen-year-old in the trash.

In the end, it turned out that this lead, like most of the others, was a hoax. I doubt that this jerk will ever know that his call resulted in our canceling our mountain search. More than nine months later we will learn that while searching the foothills near Dry Creek Canyon, we were within a mile of the "camp" where Elizabeth Smart's kidnappers were holding her. I don't know that we would have found her had we not been diverted to construction areas, but it might have shortened her abduction time from 280 days to one. No one will ever know.

Seven Choppers

Helicopters save lives. Helicopters take lives. It's a paradox.

According to *USA TODAY,* from 2000 to 2005, sixty people died in eighty-four air ambulance crashes. As difficult is this statistic is to believe, more than ten percent of the U.S. air ambulance helicopter fleet crashed during this period. If commercial airlines lost the same proportion of passenger jets as air ambulance companies lose helicopters, a large jet would crash every four days.

In the Mountain Rescue Association's "Accidents in Mountain Rescue Operations" report, Charley Shimanski says, "The most common accident that results in injuries or death in mountain rescues is the aircraft accident. In particular, accidents involving helicopters…"

Data from National Park Service rescue collected from 1925 through 2001 showed that more than half of all rescuer fatalities were the result of aviation accidents, most of them involving helicopters.

Experienced mountain rescuers know these statistics. If you die during a mountain rescue, it's much more likely to involve a helicopter than a rope. Members of Salt Lake County's team know this all too well.

On January 11, 1998, three years before I joined the team, an Airmed helicopter picked up a backcountry skier injured in an avalanche in Little Cottonwood Canyon. Moments after lifting off into deteriorating weather, the helicopter slammed into the side of the canyon. Everyone aboard—the pilot, paramedic, nurse, and patient—was killed, and in an eyeblink the mission changed from an avalanche rescue to an aviation recovery.

To an outsider, helicopters seem so convenient. Why spend precious time and enormous effort hiking in to help someone in the mountains when you can fly in? Many urban rescuers seem oblivious of the dangers of using helicopters during mountain rescues. They tend to view rock climbing and backcountry skiing as dangerous, and helicopters as safe—and reality is precisely the opposite.

The flipside to these sobering statistics is that helicopters are fast and do save lives—lots of them. I was frequently inserted using a helicopter, and most of our critically injured patients flew out on them.

Salt Lake County is unique in that one of the air ambulance companies, Life Flight, has a ship that can hoist a patient into the sky, without landing. The paramedic and patient are then flown, hanging outside the aircraft, to a safe LZ, where they can be lowered to the ground. Life Flight has well-trained flight crews and an impressive safety program; many of our patients owe their lives to Life Flight's hoist.

Difficult Decisions

The paradox that helicopters both save and take lives complicates the decision whether to use them. Consider a hiker who has fallen and fractured his leg. It will take a ground team two hours to reach him and five hours to carry him out. Alternatively, we can bring in the hoist and have him in a medical center within an hour. Which is better for the patient? Which is safer? Which is the best use of resources?

To help answer these questions, our team developed guidelines stipulating that we would use the hoist only to improve patient outcome—which was loosely defined as saving life or limb—or to improve the safety of the rescuers. I know that the second criterion sounds ironic. How can a helicopter *increase* the safety of the rescuers when, according to the statistics, they're the biggest hazard that rescuers face? The answer depends on the situation. For example, instead of traveling through terrain where avalanche or rockfall is a real risk, the rescuers may indeed be safer if inserted by helicopter.

Overreliance

Aside from safety considerations, reliance on helicopters over boot leather can also result in a weakened team. If helicopters are the backbone of rescues and a storm rolls in, if the helicopter is on another call, or if a skilled mountain pilot isn't available, the team may not be prepared for a ground rescue. Whether the team has gotten out of shape and unfamiliar with the local trails, grown a bit rusty on some of the more technical aspects of mountain travel and rescue, or just isn't up to providing patient care over long periods, overreliance on technology can undermine critical skills.

When I was running a rescue and decided to use a helicopter and leave most of the rescuers sitting in the command post, I knew they were frustrated and demoralized. It wasn't that they disagreed with most of these decisions—after all, they're team players who want what is best for the patient—but we all knew that having rescuers standing at the trailhead and saluting the helicopter as it flew overhead wasn't the best use of their time or skills.

Air Cowboys

In late 2001, seven moose were hit and killed by cars on the section of Interstate 80 that runs between Salt Lake City and Park City. With the 2002 Winter Olympics only a few months away, the Utah Department of Wildlife Resources decided it was time to "thin out the moose herd." Not your typical problem, and it didn't call for a typical solution. Bring in the air cowboys.

This crew of four modern cowboys would ride the range using a small red helicopter for some high-tech team roping. While the pilot chased the moose, the "gunner" would fire a net from a handheld device. The pilot would quickly land, and the two "muggers" would spring from the helicopter and disable the moose by hogtying three of its legs. It would then be blindfold to calm it and placed in a large bag, so it could be transported to a waiting trailer and relocated to a less populated area.

On this cold January day, the cowboys had rounded up four moose and had just dropped off one of the muggers near the shore of a frozen lake. My pager reported that a helicopter had crashed in a lake.

When we arrived, the ship was upside down in the ice with most of the cabin submerged in the frigid water. Its skids and belly were facing the sky. The flimsy helicopter had punched a circular hole in the thick ice.

Concerned about rescuers falling through the shattered ice, we dragged a small inflatable boat over the hard surface, leery of the red and white tail section of the helicopter twisting above us like a giant Calder mobile in the high-voltage power lines.

After winching the helicopter from the ice, we removed the three crew members, who had died in the crash. Although they had been submerged in the icy water, their clothes were soaked in fuel. We loaded them into the inflatable boat and dragged it to shore.

The victims had probably felt that the dangerous part of their job was while they were capturing the moose. Flying from one point to the other probably didn't seem risky, and they were no doubt taken by surprise when the tail of the helicopter hooked the powerline, flipped them, and pitched them to their death.

It's a good reminder for me that just because something feels familiar and that I've done it a lot of times—whether skiing in avalanche terrain, rescuing people in the mountains, or driving a car, there are still inherent risks, and it doesn't do to get blasé.

Heatstroke on Mount Olympus

I worked out with weights this morning for the first time in a month. With my scrawny muscles exhausted, I headed to the indoor climbing gym with my daughter. The page came as we were untying from our last climb. I've been on the team for two years and this will be my eighty-eighth callout.

```
This is a callout for S&R on a
heatstroke victim at Mount Olympus
Trail Head..5600 S Wasatch
Blvd...respond on Code One Frequency.
case#6501..eom
17:39 06/07/03.
```

Mount Olympus in summer is a textbook setting for heat-related illness. The main trail faces due west, into the hot afternoon sun. And it's steep, rising from the valley, at 4,600 feet, to the summit, at 9,026 feet, over a relatively short 3.5 miles.

"Mount O" is a tempting hike to people in the valley. For one thing, it's right there. Rising like a steep volcano, this monolithic feature is visible from just about anywhere in the Salt Lake Valley. The trailhead is impossible to miss—in fact, for many urban dwellers it may be the only trailhead they see.

Although I don't know it yet, the temperatures will exceed a hundred degrees twenty-four times this year. The high today is expected to be near ninety. On days like this, you simply cannot carry enough water up the mountain. The twelve-ounce bottle clasped in a hiker's hand is nothing in this hot, dry desert air.

My pager reports that the patient is suffering from heatstroke—something very different from heat exhaustion, although people tend to intermix the terms. Heat*stroke* is serious, a true medical emergency in which the body's ability to dissipate heat is overwhelmed. As a result, the core temperature rises rapidly to the point where tissue is destroyed and survival is in question. Heat exhaustion, though serious, is only a precursor to heatstroke. Although this call is reported as heatstroke, all we really know is that the patient is showing signs and symptoms of a heat-related illness, or hyperthermia. Dehydration will likely be a contributing factor.

Driving to the trailhead, I learn that it was a doctor who called 911 and reported that the victim was having heatstroke. That adds validity to the diagnosis.

When I arrive at the trailhead, paramedics from County Fire are already there. I see Paris Napoli, a paramedic with Station 12. I first met Paris at an advanced trauma life support class, and since then we've worked a dozen or more missions together. I'm always glad to see him at the trailhead. A firefighter who appreciates the different roles and skill sets of Fire, on one hand and Search and Rescue, on the other knows that we have strong backcountry medical skills, just as we know that his medical skills dwarf ours. Rather than wait at the trailhead for several hours, Paris decides to take his crew back to the fire station. He asks us to call him when the patient is a half hour from the road.

It takes us about an hour to gain 1,800 vertical feet over a mile and a half and reach the victim. Twenty-year-old Ashlee is lying in the dirt and looks miserable. She reminds me of what it feels like to have the flu and feel so wretched you wish you could be put out of your suffering. It's awful enough when you're at home in your own bed, let alone lying in the dirt several miles up a steep trail, in 90-degree heat.

As I put the stethoscope in my ears, my heart is still pounding hard from the hike, and I can hear it *wooh-whoosh-ing* away.

After performing an assessment to determine her medical needs, I kneel down to start an IV. Fluids will help her, but she's so dehydrated, her veins have shrunk completely out of view. As if lying in the dirt and being puke sick in the mountains weren't enough, now some guy is going to go fishing with a needle in her arm.

To Hoist or Not to Hoist

Andy Peterson takes over the medical care as I work on a transportation plan. Since this looks more like heat exhaustion than heatstroke, the situation doesn't really justify using a helicopter. Also, with no landing zone nearby, it would have to be a helicopter with a hoist.

The ground evacuation will be tricky, though, because the Mount O trail is too narrow to have rescuers on either side of the litter, so we'll need to use the wheel. And the trail is steep enough that the litter will have to be belayed by ropes the whole way down.

A rescuer is trudging in with the wheel now, but it will probably take another hour to arrive. And although a fit hiker can get to this location in under an hour, taking a patient down the trail will be slow going, probably taking four or more hours.

Andy comes back to me and says, "Steve, we've got to hoist her." I remind him that there isn't a life-or-limb threat. Gesturing toward our patient, lying in the dirt some twenty feet away and dry-heaving, Andy says, "Steve, if you were that sick, would you want to ride in a litter for several hours?" He also brings up the challenge of protecting her airway by making sure she doesn't aspirate vomit while on her back. Andy is right—bring in the helicopter. Maybe I'm looking for a way to justify it, but aspirating vomit is a life threat. Also, her condition may deteriorate, and having the team available for another call—a fifty-fifty chance on a Saturday in June— may save somebody else.

Three of the rescuers on this call are newbies, and several others have been on the team for about a year. This is a good opportunity to remind them about helicopter safety and to emphasize our protocols. To make my point, I tell the group, "Let's have only two rescuers in the crash zone when the paramedic is lowered." Then I send two of the newer members up and down the trail to make sure that "no public gets in the crash zone."

I realize that "crash zone" is a bit melodramatic, but instilling a little fear of helicopters now might save their lives someday.

With the trail blocked above and below our patient, paramedic Brian Allred is lowered from the hovering chopper to the trail. After his quick assessment, we pack the patient in the beanbag, and the ship returns and hoists our patient into the sky. She's in the air, and I am feeling good. The helicopter was the right call—a little borderline, perhaps, on whether she met the risk criteria for a hoist, but I think it was a good risk assessment.

I put the litter on my back and, still feeling pretty zippy despite the earlier workout and indoor climbing session, decide to jog down the trail. At least, I think of it as jogging, but with the weight on my back and not wanting to accelerate the wear on my aging backcountry knees, I do an old man's jog, making sure one foot doesn't leave the ground until the other has touched.

Life Flight Down

On the radio, I tell the incident commander that the patient has been hoisted and is "inbound to the command post." Now, part of sounding professional on the radio—or maybe it's just part of sounding cool—is to sound a little bit bored. Emotion is amplified by radios, and the calmest voices sound the most professional.

Knowing this, I'm surprised to hear Sgt. Thad Moore's voice on the radio, sounding anything but cool or bored:

"Life Flight crashed! Life Flight crashed! Life Flight crashed!"

We are on the mountain, with mountain rescue gear, and our rescue helicopter has crashed.

Brian, the paramedic we just handed our patient to, is now a patient himself—if he's alive. I wonder if Ashlee was still on the hoist when the ship crashed. It was my decision to hoist her. Over the coming weeks, I will have feelings of doubt about whether I made the right call.

I see a huge dust cloud come up from a ridge less than a mile from me. It looks like a cloud of smoke, but it's brown— dirt. The cloud rises several hundred feet and hangs in the air.

I consider taking the direct, over-hill-and-dale route to the crash site, but I don't want to get bogged down bushwhacking, so I decide to head down the mountain to the road, taking the normally forbidden shortcuts between switchbacks.

Now I'm running fast, no longer concerned about my knees or the litter on my back. When I reach the parking lot twenty minutes later, I put a Kojak light on the roof of my truck and barrel down the road. There isn't any traffic, since deputies have closed the road at both ends. I've been seeing a steady stream of emergency vehicles racing to the crash site near the base of the mountain. The radio has put out a steady stream of requests and reports.

Listening to the radio, I hear that Paris, who has been monitoring our radio frequency since returning to the station, has self-dispatched to the crash site.

As I near the site, I hear that an Airmed helicopter is inbound. I park my truck sideways across the road to make it clear to the pilot that the road will not have traffic and that I know he's coming.

I walk fast to the crash site, which is only a hundred feet uphill from the road. Fire suppression foam covers the ground. Paris and his crew are carrying Brian on a stretcher. He has almost every extremity splinted and wrapped in bandages. They carry him to the Airmed helicopter while the nurse, Denise Ward, is taken to a ground ambulance. I'm relieved to learn that Ashlee was transferred to a ground ambulance before the ship crashed.

Looking at the crumpled ship, I can see that the pilot, Brent Cowley, is dead. *Aw, shit!* Brent has done countless missions for Utah's search and rescue teams and has saved a lot of lives. It doesn't seem possible—or the least bit fair— that a guy who has rescued so many should die in the process.

I walk up to the helicopter and look at its twisted rotor. It is made from beer-can-thin metal that I can almost dent with my fingers. The cockpit has broken completely away and is facing the rear of the craft. The wreck is sitting in the middle of a horse corral at the base of the mountain. There is agreement that Brent managed to put the crashing bird here and, quite likely, save Brian and Denise.

Aftermath

Investigators discover that the tail rotor is missing. They will later learn that a small metal shaft, not much bigger than a flashlight, inexplicably sheared, causing the tail rotor to separate from the ship. Witnesses attending a barbecue several miles away said they saw the rotor fall off, so I take GPS and compass readings from their house and extrapolate the likely location. We begin searching for it just before dark and keep going into the night.

After the rotor search is called off for the night, I drive to the hospital for a crisis meeting. I figure there will be a dozen people standing around. I want to be there for Brent and let the Life Flight employees, our extended family, know that we appreciate what they did for our patient.

I am directed to an auditorium in the hospital, where more than a hundred people have gathered. Most of them are Life Flight employees, but there also are several deputies, firefighters, and people from Airmed.

Denise, the flight nurse, is here. She explains that after hoisting the patient, they transferred her to a ground ambulance and were returning to the original location to get GPS and altitude readings. She heard a *pop,* and the helicopter started spinning. She was trying to brace herself when the helicopter hit the ground. Once on the ground, she climbed out a hole where the front of the cockpit had broken off and spun around. Grabbing a radio, she called dispatch to ask them how to shut down the engines. She then cared for Brian, who was unresponsive.

Denise was walking around when the medics arrived. Interestingly, she would have known that based on the mechanism of injury, she should have her spine immobilized—but then, rescuers rarely follow protocol when they themselves are the patient.

The people in the auditorium are hurt, angry, and in disbelief. Only five months ago, a Life Flight helicopter crashed near the airport in heavy fog, while going to a car accident. On that call, pilot Craig Bingham and paramedic Mario Guerrero were killed, and flight nurse Stein Rosqvist sustained serious injuries. Somebody leads a prayer.

Lying in bed that night, I'm grateful that Ashlee is safe. She was the only one at risk who didn't make a decision to ride in the helicopter. The pilot, nurse, and paramedic know the risks, as did the rescuers on the ground. But I made the decision for Ashlee. If she had died or been badly hurt, the feeling of guilt, deserved or not, would have been overwhelming.

EIGHT
SLIP AND SLIDE

I'm at my daughter's soccer game when the first page arrives:

```
NEED S&R ON A FALLEN HIKER MEET AT
THE MT OLYMPUS TRAILHEAD COME 10-8 ON
CODE 1 CS 50001 CAR 606 IS IC EOM
14:23 05/15/04
```

This time I think it really *is* a fallen hiker, since not many people go rock climbing on the Mount Olympus trail. The soccer game is on the other side of the Salt Lake Valley, but I'm not too concerned—we get a lot of calls for tired or injured people on the Mount Olympus trail. This should be run-of-the-mill. I make arrangements for someone to drive my daughter home and jog leisurely from the soccer field to my truck.

On the drive to the staging area, my pager sounds again. Since it's been only a few minutes since the first call, I assume this is a duplicate page. But this one says there are *two* fallen hikers.

```
NEED S&R TO RESPOND ON 2 FALLEN
HIKERS MEET @ MT OLYMPUS TRAILHEAD
606 IS IC CS 50001 EOM
14:30 05/15/04
```

I can think of only a handful of calls that had more than one patient, and these were usually the result of a motor vehicle or aviation accident.

About ten minutes later, I get a third page, changing the meeting location from the Mount Olympus trailhead to Thousand Oaks Drive.

```
NEED THE COMMAND POST BROUGHT TO 4500
S 3815 E (THOUSAND OAKS DR) ON THE
FALLEN HIKERS IN MT OLYMPUS CS 50001
EOM
14:42 05/15/04
```

Ah, this is making more sense! Thousand Oaks Drive is in a residential area near the base of Mount Olympus's north face. The north face is a rugged quartzite wall rising some twelve hundred feet. These "hikers" are probably climbers. Once again, what started as a fairly routine call—a fallen hiker on a popular trail—now looks a lot more serious: two fallen climbers near the north face.

I will later learn that one of the people in the party of three, Nathan, called 911 on his cell phone. He told the dispatcher that both his climbing partners had fallen and were injured in the Z Couloir on Mount Olympus. The first page was sent to the rescue team as soon as the dispatcher realized the call was for an injured person on Mount Olympus. It took twenty minutes for the deputy on scene to learn that it was in the Z Couloir and have the information repaged.

The radio continues to provide updates as I drive. The latest reports say that we have two female patients who both fell down an icy couloir—a steep, narrow gorge in a mountainside. A helicopter has been called to assess the situation. The flight crew reports that they can see a patient, who appears unresponsive, about a thousand feet up the snow couloir. They tell us that the rock walls are too steep to bring in a rescuer on the hoist.

As I drive past the Mount Olympus trailhead, I see a parked fire engine. I'm guessing they haven't been told of the new location—or I'm on a wild goose chase.

I am one of the first rescuers to arrive at the command post, which is parked in a cul-de-sac in ritzy neighborhood. I'm a little surprised that there aren't more rescuers here, because it took me a while to get here from the soccer field. Others must be coming from across the valley or down from playing in the mountains on this beautiful Saturday. It's been almost forty-five minutes since the initial page.

I take the trauma kit, oxygen rig, and an ice ax from the back of my truck, then lace on my plastic ice-climbing boots and stuff my crampons into my backpack. The spikes are sharp, and like just anyone who has spent a lot of time on crampons, I've accidentally hooked one in a pant leg on several occasions, and once fallen as a result.

The north face is a formidable-looking wall: sheer, with its upper ridge outlined against the sky. It looks as if the mountain had been sliced in half with a giant knife. Tens of thousands of people drive by it each day, but I think few actually see it, and only a handful ever climb the vast expanse of the north face.

Dan Smith and I work our way onto the trail at the end of the cul-de-sac. Now that another new house has been built and the owner has put up a chain-link fence, the trail has been forced uphill and runs along the fence—now the boundary between the private and public property. From there it zigzags up the hillside to the base of the snow-filled couloir. When we reach it, we stop to attach our crampons to our boots. Carrying ice axes for safety, we begin hiking up the steepening snow. On steeper ice, these tools, along with our crampons, would allow us to climb vertically. Today we are using them to prevent us from slipping and to stop us if we should fall.

My pager goes off again.

```
NEED MORE S&R UNITS TO ASSIST ON THE
FALLEN HIKERS @ MT OLYMPUS MEET AT
THE TRAILHEAD AND BRING ICE PACKS 610
IC CS 50001 EOM
15:16 05/15/04
```

Although I only glance at my pager to make sure we don't have another call, I grin at the request to "bring ice packs." I'm sure they intended to tell the rescuers to bring ice-climbing gear, but I wonder if anyone is going to show up with those little ice packs that you might press on your forehead to relieve a headache.

After hiking up the snowfield for about twenty minutes, we meet the first victim. She is being helped down by bystanders. Her name is Elissa, and she is twenty-eight years old.

"Command Post, Team One has arrived at the first victim," I say over the radio. Since the first page, it has taken almost an hour and a half to reach Elissa.

I introduce myself and ask if she's okay. She tells me she is fine other than an injured right arm, but that her friend, farther up the snowfield, is hurt worse.

Seeing blood-soaked bandaging around her right elbow, I expose it and discover what may be an open fracture.

Although many people believe a "fractured" bone is different from a "broken" bone, in medical terms they are synonyms. Without X-ray vision, we can only go with the signs and symptoms. Sometimes the fractures are obvious, like a climber I saw who lost control while rappelling and fell twenty feet. The bones in his lower leg were so shattered that his foot and ankle were attached to his leg by his calf muscle alone.

If the skin over a fracture is cut, it's called an "open fracture," regardless of whether the bone is protruding from the wound or whether skin was actually cut by the bone. In any case, the prehospital care is the same: control the bleeding and stabilize the bones.

Three bystanders have been helping Elissa down: one on either side and one below her, who has been walking backward down the slope. I thank them for helping and tell them we would like Elissa to sit down. They seem a little surprised, since they have been doing a good job of getting her off the mountain, but we don't want her to slip—we can't afford another fall.

Elissa explains that they just climbed the "West Slabs" on the north face, having started at five a.m. At 5.5 on the Yosemite decimal scale, the climbing is not very difficult, but it's long—twelve hundred feet, which is ten to twelve roped pitches. If you fall during the climb, you're going all the way to the bottom.

I will later learn that after summiting at ten a.m., the three climbers made ten rappels to get back down to the couloir. All three climbers had ice axes, but Elissa and Cheri were inexperienced at using them.

They were glissading—French for "gliding," down the steep snow and ice couloir. You can glissade either by sitting on the snow and sliding feetfirst down the hill, as if going down a playground slide, or, if the snow is firm enough, by standing up and "boot skiing"—basically, skiing without the skis. Glissading requires the ability to stop yourself quickly using an ice ax, with a technique called "self-arrest," because it's easy to pick up speed and lose control. If you get going *too* fast, not even the ice ax will stop you.

Nathan glissaded a few hundred feet and stopped to wait for the others. At first he couldn't see his wife, Cheri. Then, hearing a scream, he saw her come flying down the slope. Nathan tried to block her with his body, but she had too much momentum and knocked him off his feet. She tumbled another hundred feet, bouncing off the walls, until she came to a sudden stop, wedged between the snow and rock.

Nathan will later tell reporters, "I ran down and heard her gasping for air. I thought she had punctured a lung, but her mouth was full of snow and her face was pinned against the snow bank." The skills Nathan learned in a wilderness first responder course may have saved her life when he cleared her airway, stabilized her spine, and added clothing to keep her warm.

When I later hear of his actions, it will bring back memories of an injured person who slid into a tree in the back-country. A young man who had taken the ski patrol's outdoor emergency care course happened to ski up moments after the accident. Seeing that she was struggling to breathe, he knelt down and repositioned her head, and she took a deep breath. I don't know if this young man knows it, but I think he saved her life. He wasn't thanked—we were too busy. I hope he reads this book.

As Nathan was treating Cheri, Elissa stopped uphill to pickup Cheri's ice ax. As she reached for it, she slipped, slid, and smacked into the rock wall.

I use my radio to call down Elissa's vital signs and medical condition as a "possible open fracture of her right elbow." Dan then takes over Elissa's medical care as I continue up the couloir to our second patient, Cheri, who is a few hundred feet above us. When I arrive, I find a woman who appears to be in her early twenties, wedged in a crack between the ice and a rock wall.

During the spring, the sun heats the rocks, slowly melting the snow, which pulls away from the wall. This creates a gap, called a moat, between the snow and the rock wall.

Cheri is really wedged into this crack, which continues downward another six feet. It sort of looks as if a snow-and-rock mouth were trying to swallow her.

"Command Post, this is Nine-oh-one. I have arrived at the second victim." It is amazing how our roles can shift so quickly. Less than two hours ago, I was a father supporting his daughter at her soccer game, and now I'm on rock and ice, involved in a mountain emergency. Nathan and crew probably feel a similar shift. A few hours ago they were coming off an exhilarating climb, and now things feel suddenly very different.

There are three or four people standing around the victim. They were hiking below when they heard Nathan call for help. Most of them are wearing shorts and tennis shoes and are shirtless; all seem oblivious of the danger they are in. Yes, you can walk on this snow slope. But as Elissa and Cheri demonstrated, you are always only a fall away from potentially life-threatening injuries. Yet even while standing here, looking at Cheri and seeing the serious results of her fall, they exude that teenage sense of immortality. I am concerned that a little slip by any of them will create a third patient, maybe more. I feel a bit like a parent watching a happy toddler walk down a sidewalk by a busy street.

"Command Post, Nine-oh-one."

"Go ahead."

"Due to the conditions, I don't want any rescuers on the snow unless they have an ice ax."

I know my decision will not be popular—even rescuers feel immortal at times. The conditions are not that hazardous; it is easy to walk on the snow. But it's also easy to fall, and the consequences can be deadly.

I begin my medical assessment: "Hi, my name is Steve, and I'm with the Salt Lake County Sheriff's Search and Rescue Team. Can I help you?" The question always seems absurd—of *course* you can bloody well help me! But the answer does a couple of things: it gives me permission to treat the patient and also helps me understand her level of responsiveness.

She asks me what happened. I tell her that she fell, and I ask her a series of standard questions as I begin a head-to-toe assessment.

"Do you have any allergies? Are you taking any medicines? Do you have any preexisting medical conditions or are you seeing a doctor? What's the last thing you ate? What happened?" These questions are based on SAMPLE, an acronym that reminds me to ask about symptoms, allergies, medications, preexisting medical conditions, last intake and output, and the events leading up to her situation.

When I ask her age, she tells me she is twenty-four years old. Nathan, who is watching over my shoulder with obvious concern, tells me she is actually twenty-eight. I ask her several more questions, including about Nathan. She says she knows him but can't remember his name. This inability to recall her own husband's name, along with her apparent lack of concern over such a serious lapse, is a clear sign of a head injury.

Although I can't access most of her body because of her position in the moat, she complains of pain only when I palpate her jaw. It looks as though she may have missing or broken teeth. I make sure there aren't any tooth fragments in her mouth that she might choke on. Her jaw is very tender and I suspect a fracture.

She continues to ask me what happened. Repetitive questioning—another sign of a head injury. My mind momentarily drifts to my patient with the most memorable repetitive question. A doctor who had injured her head while skiing kept asking, "Am I doing repetitive questioning?" Each time I answered yes, she'd say, "Crap, I don't want to have a head injury." And a few minutes later she would ask the same question, verbatim.

A trickle of dried blood extends from Cheri's left ear. I think it's from a cut near her ear, but I can't tell. I'm hoping it isn't a sign of a fractured skull, which can cause blood or cerebrospinal fluid to leak from the ears. I have a feeling of déjà vu—a week ago today we treated a fallen climber who had blood running from his ear due to a skull fracture.

I want to put her on high-flow oxygen using the face mask, but at that rate, the bottle will last less than twenty minutes. Instead, I opt for a nasal cannula, which will give Cheri almost an hour of oxygen, though it at a much lower rate of flow. I pull out the tank and crack the valve, which releases a loud blast of air that echoes off the rock walls, and connect the regulator and gently fit the prongs of the cannula into her nostrils.

Rechecking her vitals, I see that her pulse and breath rate have increased since I arrived—not good signs, for they imply that even with the supplemental oxygen, her body is struggling to keep itself oxygenated.

I radio down for two more oxygen tanks. If I can get tanks staged on the way out, I can shift to high-flow oxygen. Over-anxious, perhaps—I feel a hole in the pit of my stomach. I'm concerned she is going to die, as my brain-injured fallen climber did last Saturday.

It's obvious that Cheri is seriously injured. I'm concerned about her head injury, since these have a way of deteriorating quickly. The skull is a sturdy, rigid container that protects the brain. But once the brain is injured, that same rigidity can be a detriment, trapping blood inside and increasing pressure on the brain.

She is also complaining of back pain, but with her so squished into the rock-and-snow moat, I can't palpate her spine. We have to assume she has a spinal injury.

Her breathing, although at a reasonable rate, sounds labored. I don't know if it's because she is jammed into a crack and can't fill her lungs, or because there's a lung injury. I suspect the latter.

Although I am right next to her, I can't really say that I have accessed her. We *have* to get her out of this moat.

Rescuer Amy Fisher arrives. Amy was my training officer when I joined the team, and is one of only two women firefighters in Murray City. If they can find any more like her, they should sign them up. She suggests that we tie webbing around Cheri's waist so she won't sink any deeper into the moat. It's a good call, but she is so jammed into the crack, it isn't easy to thread the rope around her body. Amy works the webbing around Cheri and then climbs about ten feet up the rock wall, finds a place where she can brace herself firmly in the wall, and assumes a belay position. Amy is now a human anchor.

My pager vibrates as more rescuers are summoned:

```
NEED MORE UNITS TO ASSIST WITH THE
FALLEN HIKERS IN MT OLYMPUS MEET AT
THE TRAILHEAD COME 10-8 ON CODE ONE
CS 50001 601 IC EOM
15:43 05/15/04
```

I cover Cheri with the heavy-duty space blanket from my backpack. This should help keep her warm and protect her from snow during our upcoming excavation project.

Nathan wants to help, but when I ask him to step back so we have more room to work, he does so willingly—that's the best help we can get in this moment. Often, well-meaning people want to help, but their efforts can actually interfere with our helping their friend or family member, as happened with the woman who got herself lost while searching for her missing daughter on Mount Olympus. I've also been surprised at how willing some people are to step back and let us do our work—like the ER doc who didn't need to be asked to stand back while we assessed and packaged his wife after a backcountry skiing accident.

I swing my ax to chip away at the ice that encases Cheri against the rock wall. What was spring snow has turned to ice after repeatedly compressing, melting, and refreezing. It's a tedious process, with each swing breaking loose just a few ice cube–size chunks. I first cut a three-inch trough about a foot out from Cheri and then start breaking out the larger chunks. After fifteen minutes of chopping, my arm is so fatigued that I'm no longer effective. At that point, Dan arrives and takes over. It's hard work, and painstaking, too—we have to keep the sharp ax from glancing off the ice and into Cheri, and we also have to avoid cutting the tubing that connects her to the oxygen.

It takes us forty-five minutes to free her from the icy maw. Then, using the snow and ice that we've excavated, we build a small ramp below her. We need to get her out of the moat, but it's imperative that we limit movement of her spine during the process.

Other rescuers have been arriving with additional gear. Some are working on building the ramp; others are flaking out ropes, setting up the litter, and helping with medical gear.

We fit the patient with a cervical collar and roll out the full-body vacuum splint. The beanbag is equipped with a big gray pump that looks like a bicycle pump from a Warner Brothers cartoon. Using the pump, we evacuate some of the air from the beanbag to stiffen it. Then, on my three count, several rescuers lift Cheri a few inches while keeping her spine as stable as possible. After a rescuer slides the rigid beanbag under her, we release the vacuum, wrap the bag around her, and revacuum the bag to make it rigid. Amy is still in position as belayer.

Alan Erdahl has been making a "deadman" anchor in the snow. In his nearly forty years on the team, Alan has seen it evolve from Jeep posse to a finely tuned mountain rescue team, and he has done a stellar job of adapting to the changes. After responding to almost two thousand callouts, he knows the Salt Lake County backcountry like no one else.

Alan creates the anchor by using an ice ax to cut a horizontal trench in the ice, about eighteen inches deep and six inches wide. Then, from the center of the trench, he cuts a narrower perpendicular trench to form a T running down the fall line of the hill, with the first trench forming the top of the T. He ties webbing around two ice axes and sets the axes lengthwise in the horizontal trench, with their picks sunk into the hardened snow, and runs the webbing downhill in the vertical trench. After burying the axes with snow and ice, he stomps down on the loose material. The result is a loop of yellow webbing sticking out of the dirty snow, connected to a bombproof anchor. After connecting a brake bar rack to the webbing, he connects a two-hundred-foot rope to the rack.

Mike Brehm and Amy tie themselves into either side of the litter so they can help guide it down the mountain. Then, with the rope tied to the head of the litter, Alan lowers Cheri feetfirst down the couloir.

Alan uses his radio to keep the litter tenders posted on the amount of rope remaining: "Forty feet...twenty feet...ten feet... Stop!" Then several rescuers hold the litter still so Cheri can come off belay while Alan connects a second two-hundred-foot rope to the end of the first rope and resumes lowering.

After lowering the threesome another two hundred feet, we repeat the process, this time connecting a 165-foot climbing rope—it's what we have left.

At this point, with Cheri, Mike, and Amy on the end of more than 500 feet of rope, the stretchy rope acts like a long, very strong rubber band. Though Alan is trying to maintain steady tension on the rope, the team goes from being stopped to moving too quickly, as the rope stretches and relaxes—boing, boing, boing.

As Alan nears the end of our third and last rope, we switch to a new anchor, which Darren Westerfield and a bystander—it may have been Nathan—set up in rock. We move the rope to this new, much more easily built anchor and continue lowering.

The steep rock walls above Cheri's original location prevented the helicopter from safely hoisting her, but now that she is near the bottom of the couloir where it opens up, the helicopter can get in. The flight crew has already rigged for a hoist and has been hovering in the distance as we lowered Cheri.

The hoisting happens quickly. The medic is lowered from the sky and unhooks from his cable. Rescuers help load Cheri, already packaged in the full-body vacuum splint, into the hammocklike sling. The helicopter returns and lowers the hook into the medic's hand. And with a wave of his arm, he and Cheri rise into the sky. It's a beautiful sight and, as always when we hand a patient off, an enormous relief.

Once Cheri is airborne, we hike back up to Elissa, who has been waiting patiently all this time. We connect the rope to her and belay her as she walks, with rescuers supporting her, down the snow-and-ice surface. When she reaches the end of the couloir, we take her off the rope, and rescuers help her walk down the switchbacking trail to the ambulance waiting in the cul-de-sac.

I will later learn that Cheri had a concussion, torn shoulder ligaments, a broken rib, and a lung injury. Her jaw was broken in three places and required surgery. I never found out whether Elissa's arm was fractured.

Life Flight hoists Cheri to safety
(Visit MountainResponder.com/Photos and enter Story Code MOG.)

Nine
Waterworks

Water—it's strange stuff. Heavier and denser than a hardwood floor, but you can't walk on it. It can be as clear as air, but you can't breathe it. It can support a fifty thousand–ton cargo ship but will swallow a fifty-pound child.

Water in one form or another played a part in many of the rescues I participated in. I've seen snowflakes released en masse to bury skiers in avalanches, retrieved climbers from waterfalls, seen swiftwater drown the unwary or inebriated, and found hikers who slipped on a patch of the frozen stuff and slid into rocks or over cliffs.

Swiftwater Missions

Most years saw a handful of swiftwater missions. These were almost always recoveries rather than rescues—when someone falls into fast-moving water, they either get out fast and don't need our assistance, or drown. There just isn't much leeway between the two outcomes.

Child in River

I've been on the team for only a few months, and my daughter and I are driving home from her soccer practice. I'm guzzling water to make sure I am well hydrated for the team's fitness test, which starts in less than an hour. Being new on the team, I'm a little nervous about the timed test, which requires us to carry a weighted backpack up a trail that rises fifteen hundred feet in less than two miles. The time requirement is generous enough, but I feel the unspoken need to make a strong showing.

We are still a few miles from home when my pager goes off. I expect it to be an informational page about the fitness test, but no such luck:

```
MURRAY PD IS REQUEST S/R TO ASSIST
THEM AT MURRAY PARK 475 E 5335 S FOR
A 7 YOA THAT HAS FALLEN IN THE CREEK
THIS IS ON CD1 EOM.
05/17/01
```

This is only my second callout, and the message that a seven-year-old has fallen into the creek fires me with a hefty dose of adrenaline. Over the coming years, I'll get better at controlling my response.

After dropping my daughter off, I have to drive through ten miles of city traffic to get to Murray Park. On the way, I see a white sheriff's office pickup with lights flashing and siren wailing, towing a trailer with two inflatable boats. The spectacle draws curious stares.

A radio update redirects us to a bridge several miles downstream from the initial staging area. When I arrive, I see a sheriff's office SUV pulling a trailer loaded with twenty-foot lengths of inch-and-a-half metal pipe. Teams of two rescuers are connecting sections of pipe using steel pins. Other rescuers then carry the pipes to the upstream side of the bridge, where they are stood upright on the river bottom, leaned against the bridge, and secured with rope. This is repeated so there will be a pipe every ten inches across the fifty-foot-wide river.

The bridge is a hive of activity as rescuers work feverishly to get the grid up. I join the ones connecting the pipes.

Working with big leather gloves, hard hat, and hammer, I feel like an apprentice pipe fitter.

Our efforts result in a vertical strainer designed to catch our missing seven-year-old, or so I initially think. I soon realize it is intended to catch the *body* of a seven-year-old.

The obvious first priority on a swiftwater call is to deploy hasty teams immediately to search for a possible survivor. The not-so-obvious second priority is to get a grid set up to limit the size of the search area. The grid, as with the boats, usually arrives with lights and sirens.

Over the coming years, I will learn that deciding where to install the body-catching grid is more science than art. By multiplying the elapsed time since the person was last seen by the speed of the river, we can estimate the maximum distance the body could have traveled. The grid is then constructed on a bridge that is at least that distance downstream.

It may seem pessimistic to be building a grid almost immediately after we get the report of a missing person, but if we don't catch the body now, it could float many miles and not be found for months, if ever.

As soon as the vertical grid is up, debris begins collecting on the vertical pipes. It consists primarily of branches, but there is also a lot of wood and trash. Within an hour, the collected debris starts to overload the grid and rescuers are sent out in a tethered boat to peel off the debris.

As the boat crew is clearing the grid, a radio call comes in telling us we can disassemble it. Word is, the Murray police do not feel the initial report was credible, and after several hours without a report of a missing child, they are confident that it's a false alarm. The dismantling isn't as rewarding as the construction, but the manual labor feels good.

Moments later, the 911 dispatcher reports that they just received a report of a missing child. We stop the disassembly and wait for further instructions. Within ten minutes, that child is found at a neighbor's house, and we go back to work.

After dismantling the grid and storing it on the trailer, a deputy tells us the rest of the story. Apparently a boy around ten years old told someone that he was playing by the river in Murray Park when he "saw a kid" whom he guessed to be about seven years old, in the river. The seven-year-old was grasping a log and calling for help. The boy held out a stick for the seven-year-old, but the youngster wasn't able to hold on. As the stick slipped out of his hands, the child in the water said, "Thank you," to his would-be rescuer. I guess the thank-you was more than the Murray Police could swallow, and the boy eventually confessed to making the whole thing up.

In half an hour, I went from feeling that we might save a child to feeling like strangling one.

Three in the River

Five years and close to two hundred calls later, I've gotten fairly comfortable with swiftwater calls when this shocker comes in:

```
Search & Rescue Callout: 5200 E BCC
(Storm Mtn), report of 3 people in
the river, swift water, unk status.
Respond to Tacl, C# 43645 EOM
21:20 05/20/06
```

That someone is in the river, almost certainly drowned, doesn't shock me. But *three* people in the river? This looks like a full-blown disaster.

When I arrive on scene, a dozen or more firefighters from two agencies are already searching a section of Big Cotton-wood Creek—raging with spring snowmelt—where it drops two hundred vertical feet in less than half a mile.

I am briefed by a lieutenant who says that three teenagers, a boy and two girls, were going rock climbing at a popular beginners' crag when they fell into the river. I'm familiar with the area. To get to the climbs, you have to cross the river on two ropes—one for your feet and one for your hands. It's a bit like walking a slack and wobbly tightrope, but with an equally slack and wobbly handrail to hang on to.

I never hear the full details of what happened, only enough for us to do our job and to leave me wondering. Apparently, the three teenagers told their parents they were going to a movie at a local mall. Instead, they were going rock climbing.

As they were crossing on the rope, one of the girls fell in and was instantly whisked downstream in the torrent. The other girl, panicked at seeing her friend swept away, also fell in. The boy jumped in and tried to grab the second girl, who fell in near the edge and was holding on to a rock or root. He couldn't secure her, and they both went pelting downstream, at the mercy of the current. Somehow, the boy and the second girl ended up on the far shore, alive—probably spit out by the river. The first girl was gone.

Since I don't know the full story, it's hard to understand how so many mistakes could happen. Apparently, they didn't clip into the ropes when they were crossing—a huge mistake. And were they also all crossing at the same time? It just doesn't make sense.

Quickly spotting the lucky girl who got spit out on the other side of the river, rescuers cross on the same suspension ropes that the kids fell from—but with one key difference: the rescuers clip themselves into the rope, so they can't fall in and themselves become part of the problem. Making their way carefully down the far bank, they soon reach her. She's cold and getting hypothermic, but still shivering. Emotionally, she is even more numbed over the disappearance of her friend.

Reaching her was the easy part—now we have to get her back. This means stringing a highline across the river. Though the existing traverse is only a few hundred feet upstream from her, its ropes are of unknown strength and condition, the bank is steep, and the stream is raging—even if she weren't hypothermic, it would be too treacherous.

We string the highline directly across to her position, perhaps a hundred yards downstream from where she fell in. First, we toss a throw-bag, which spills out a lightweight rope as it arcs across the river. Alan Erdahl and Tom Moyer, who crossed the river on the old ropes to reach the victim, catch the bag. Next, we tie a rescue rope to our end of the throw-bag rope, and they pull this much stouter rope across the river. Other rescuers on our side keep some tension in the rope, so it won't dip into the torrent and be ripped from their hands. Tom then ties their end to a tree, we tie the other end to a parked SUV, and rescuers pull it taut. With the pulley-drawn litter and two haul lines attached, Alan and Tom package the girl in a beanbag, and we trolley her across the river.

As this is going on, dozens of rescuers, firefighters, and deputies are searching downstream for the still missing girl. Teams of two rescuers are dropped off every half mile to search the creek banks, while others cross the rope bridge or a footbridge farther downstream to search from the opposite side. These initial hasty teams move rapidly along the river-banks, shining lights and calling the missing girl's name. When she isn't found during the initial search, a grid is installed downstream while the teams are redeployed for a more detailed search. Firefighters use handheld infrared devices to search for the heat signature of a body, and a helicopter uses the FLIRS to comb the area from the air, looking for body heat.

Rescuer Keith Sauter is shining his flashlight on a snag in midstream when he catches a flash of something under the surface. Studying it, he realizes he is looking at the victim's foot. It's remarkable that he was able to spot it under the water.

With dozens of onlookers from several agencies, we send a tethered rescuer into the water to tie webbing onto the submerged victim so she can be removed. As we fish the fifteen-year-old's body out of the water, my fifteen-year-old is at home, asleep. I wonder what I can tell her to keep her from making a mistake like this.

At four in the morning, the final page goes out:

```
CANCELLATION on Search & Rescue
callout (updated info): 5200 E BCC
(Storm Mtn), 3 vict's - 15yof 1085
echo, other 2 bravo. C# 43656 EOM
04:44 05/21/06
```

The 10-85s give us the patients' condition, with "echo" being the obvious fatality and "bravo" signifying that the two others do not have critical injuries but do require medical attention

Alan crosses the river on the precarious ropes
(Visit MountainResponder.com/Photos and enter Story Code TIR.)

Photograph by Darren Hunsaker

Body Searches

The typical challenge in a swiftwater callout is finding the body. It is usually trapped in a snag or pinned in rocks, but it can also be caught in a Maytag-like turbulence, holding and tumbling any object that drifts into its ambit, until Mother Nature decides to release it.

A secondary challenge is to make sure no rescuers are hurt or killed while searching for someone who is already dead—risk versus rewards is an ongoing calculus in any rescue or recovery. Although everyone on the team is trained as a swiftwater technician (and retrained each year), the very fact that we are searching for a body in fast-moving water is object proof of the danger.

Swimming and Hooking

In 2002 our team spent seven days searching for a four-year-old who disappeared while camping by a raging creek. Note that the combination of "four-year-old," "camping," and "raging creek" should trigger some subtle alarm bells—many rescue situations are avoidable with a bit of forethought.

For the first few days, wearing surface dry suits, we took turns in the frigid snowmelt, thirty minutes in and thirty out, trying to stay warm. I remember feeling under rocks and thinking how horrifying it would be to find a small child's drowned body with my hands.

We also used three-pronged grappling hooks on long metal poles, scraping about in the crevices between submerged rocks to search for the body. As we were searching, the family, deciding that we weren't doing enough to find their boy, called out their church "ward," and soon had hundreds of inexperienced people searching the river. Luckily, despite their complete lack of training and the absence of rescuers positioned on the banks with throw bags should anyone fall in, there were no additional fatalities, and they did end up finding him. It made us look bad, and the family beat us up with letters to the newspapers saying we weren't doing enough—a charge that seemed to take on some legitimacy when they found the body. But in this sort of public finger-pointing, key elements are often obscured, such as the fact that we could not risk additional lives while searching for a known fatality.

A rescuer searches the raging river
(Visit MountainResponder.com/Photos and enter Story Code SWS.)

Once, while searching for a missing man in a waterfall, a fellow rescuer called me to the side. "Look at this," he said, pointing at what looked like stringy grass on his grappling hook. I rolled it between my fingers. It sure felt like hair, but we had been told that the missing man's hair was short. I called the command post and asked, "Can you confirm that the missing person's hair is short?" They called back a few minutes later and said he had short hair—mostly. It was actually a "mullet," with a long tail in the back. We had indeed found our guy.

Man Overboard

Not all our water callouts involved moving water. One time the 911 dispatcher got a call something on the order of "Help! I'm on a sailboat on the Great Salt Lake, in a squall, and the skipper was just knocked overboard! I'm the only person on the boat, and I don't know how to sail!"

Now, that was a first for all of us. There were two people who needed help: the woman on the uncrewed sailboat, and the skipper, floating out there somewhere on 1,700 square miles of lake. Some of my teammates went on the Park Service boat to wrangle the wayward sailboat, and the others took a Zodiac boat into the stormy chop to search for the skipper. The guy made it, thanks to his own resourcefulness. Finding a small buoy, he cut the mooring rope with his pocket knife and used the buoy as a float until he drifted to shore.

TEN
JUNE SNOW

On the evening of June 5, I get a call from the lieutenant in charge of Search and Rescue. He was contacted by a neighboring county; they have an overdue hiker missing in steep, cliffy terrain. They are requesting six technical rope rescuers to help them the next morning. I send a page to the team.

```
CALLOUT: Six-person technical team
needed to assist Utah County
tomorrow. Call Steve if you can help.
EOM
20:31 06/05/07
```

I don't get many responses. It can be tough getting people to show up at these preannounced callouts. First, this is in a different county, so the drive alone will add several hours. And second, the fact that they don't need us until tomorrow means the situation probably isn't all that urgent. That doesn't mean my teammates are unwilling to help, but removing the sense of urgency makes them less inclined to leave their paying jobs to volunteer in another county.

After a few more pages and phone calls, I round up five volunteers. We drive to Utah County the next morning.

When we arrive, I walk into the command post and get a briefing by Sgt. Tom Hodgson. He tells me that a man, his wife, and their young child went for a day hike yesterday. The man wanted to hike a little farther, so his wife and child returned to the car and drove home. She agreed to pick him up farther down the trail in a few hours. That evening, after he failed to meet at the rendezvous, she stopped a deputy who was on patrol. Utah County called out its search and rescue team. As he is explaining the situation, Sergeant Hodgson also tells me they are coming off a busy weekend, with numerous callouts in the past few days.

It's June, but northern Utah was hit with a late-season snowstorm that dropped about a foot of wet snow in the mountains. The snow is melting quickly, but it's still cold and the ground is snow covered. The mountains are enveloped in clouds, and sleet is falling as we begin hiking up a muddy trail.

We aren't five minutes from the command post when a few of my teammates take a wrong turn. It's a little embarrassing to have to use our radios to find each other so early in the call, but it's easy to get lost in unfamiliar, fog-blanketed terrain. After fifteen minutes of hiking, we reach an open meadow.

At the meadow, we meet with several Utah County rescuers, who point out where they searched and where they think the victim might be. To us, it seems as if they are simply pointing into the clouds—we can't see a thing. Occasionally a small hole opens up in the low clouds, and we get a glimpse of a dark rock wall striped with horizontal bands of snow.

The tired rescuers tell us the missing man is wearing a T-shirt, running shorts, and tennis shoes. The weather was nice when he left, and he obviously wasn't expecting the June snowstorm.

Our team of five hikes up a loose scree slope. The unstable rocks are covered with a few inches of wet snow, which makes them doubly tricky. As we push through brush thickets between clearings on the hillside, the wet vegetation quickly soaks the outer layer of our clothing.

We reach the rock wall and look for a route onto it. A couple of two-person teams from Utah County are still searching along the base of the wall, calling out the missing man's name. Earlier that morning, they heard a voice calling from high on the cliffs, but they no longer hear it and can't get a response. The discussion turns to whether climbers can traverse the vertical wall, into the clouds, to reach the location where the voice was last heard.

We take a closer look at the terrain. The wall is at least two hundred feet high, with narrow horizontal "shelves" about four feet wide running along it every fifty to seventy-five feet. The shelves, covered in snow, stand out as horizontal white lines across the dark rock.

We conclude that it would be possible to traverse the wall via one of the narrow shelves, but with a fair degree of risk. The consensus is that it isn't worth the risk, since we don't even know where the victim is or, more important, whether he's even still alive.

We also know that the tricky acoustics of the amphitheaterlike walls could have allowed the voice they heard early this morning to travel a long distance. We are concerned that we might travel two hundred feet along the wall only to discover he's still eight hundred feet out, over terrain that can't be crossed.

Using the LAST acronym, we would be attempting to access the victim before locating him. Yes, sometimes locating a missing person requires technical climbing, but this time the risk-versus-reward tally just doesn't add up.

Over the radio, Utah County Team Leader Chris Johnson (a.k.a. C. J.) and Sergeant Hodgson are discussing how to proceed. They are in the unenviable position of deciding whether to call off the search. Ultimately, they decide that with no sign that the victim is alive, the rescuers' safety must come first. Their rescuers have had little sleep over the past few days, and two rescuers have already been treated for hypothermia today. One rescuer was taken to the hospital in an ambulance, and the other recovered after being rewarmed and getting an IV.

The rescue adage is that you risk a lot to save a lot and risk a little to save a little. In this case, they would be risking a lot with a low probability of a save.

I think C. J. is rattled by having to make this recommendation to his command post—none of us, including him, wants to give up or delay the rescue. We verbally acknowledge that we may be leaving the victim to die. That's a tough one for people who are trained to rescue others. All of us reluctantly agree that it's the right call. It is interesting to be an outsider and watch another team face this difficult decision.

We also discuss the option of driving to the ridgeline above the cliffs and searching for him from above. That's the next best plan, but we know that searchers from above will have a slim chance of coming down the correct couloir. And searching from above will require rescuers to rappel down the snow-covered rocky cliffs. It is approaching noon, and if we search from above, it's very unlikely that we'll find him before nightfall.

If we can't find him today, it will mean that somebody in shorts and T-shirt will have to survive on a cliff, in the rain and snow, for two days without food, water, or shelter. At this point, survival seems doubtful. We are trying to accept that he is probably already dead and, if not, soon will be.

As we are walking back down the scree slope, our heads hanging forward, one of the Utah County rescuers tells us over the radio that they hear a voice. Shortly after that, the clouds part long enough for a rescuer to spot the victim. Tenacity, combined with a little cooperation from the weather gods, have resulted in the victim being located. This changes everything.

Utah County's rescuers, past exhaustion, are not up for the climb. I decide that Casey Sullivan, one of our team's strongest climbers, and I will go in. I see the disappointment in Jared's eyes and hear the grumpiness in Keith's voice, but it's part of my job to expose as few of us as possible to the hazard. In the back of my mind, I also know that there's an outside chance I will need these men to rescue us.

The footing on the shelf is iffy: loose rock and dust that has eroded, over eons, from the wall above, without being disturbed by humans. The wet dust creates a sort of dark, greasy mud that coats the wobbly rocks, and the whole unstable conglomerate is covered in snow.

Keith ties webbing around a tree to belay Casey as he heads out onto the shelf. Casey has to traverse about 150 feet along the wall before there is any form of protection: a puny fir tree, not much taller than he is, that has somehow managed to grow out of the inhospitable rock.

A challenge with belaying on a horizontal traverse like this is that until the climber can attach the rope to an anchor, falling will result in his doing a giant pendulum swing, below and beyond the point of his last good protection. If Casey falls when he's 150 feet out from Keith, he will not only drop that 150 feet vertically, he will travel almost 250 feet in a broad, swinging, *fast-moving* arc.

When Casey reaches the small tree, he ties webbing around it, clips in, and belays me across the wall. I will be in a safer situation than Casey, because Keith is also belaying me from behind. If I fall, I won't pendulum—at least on this first pitch where Keith can back me up.

I step carefully into the snowy mess, weighting each foot to make sure I won't slip, then move then other foot. The footing feels decent, and I never really feel like I will fall, but the potential consequences of a fall are sobering.

When I reach Casey, I tie into the tree and prepare to belay him. He again advances about 150 feet, until he can anchor to another tree. When it's my turn to follow, Keith gives me an opposing belay for about fifty feet, until the end of his rope comes through his belay device and he is forced to let go.

We do this three times, traveling more than four hundred feet along the narrow shelf. The process is time consuming, because we must advance one at a time, each time switching roles between climber and belayer, and managing a spaghetti pile of wet rope throughout the traverse.

As we advance along the shelf, the people in the command post realize that we might need additional rescuers, either to help with our patient once we reach him, or—and I'm glad I don't know the potential situations they are considering—to rescue us. A page for more help goes out to the Salt Lake County team.

```
SEARCH AND RESCUE CALL OUT ADDITIONAL
UNITS NEEDED FOR UT CO RESCUE, IF
AVAIL CONT 600 ASAP 555-1212,EOM
13:19 06/06/07
```

The urgency of this message flushes out a few more rescuers.

Before Casey belays me across the third pitch of our traverse, he calls over the cliff edge and hears a faint reply—it's our guy! Because of the distance, they can't communicate very well, but Casey tells him to stay there, shouting, "Don't move! Okay? Don't move!" The until-now missing person obviously is not interested in doing much moving, but Casey wants to make sure that, in a hypothermic stupor, he doesn't try climbing and kill himself before we reach him.

When I reach Casey, we're still almost two hundred feet above the victim. But our longest rope is two hundred feet long, which means that to rappel down to him, we would have to tie one end of the rope to a tree—and leave the rope on the mountain. That isn't an option, since we will almost certainly need that rope to reach the ground, which is still hundreds of feet below us.

Instead, we create an anchor by tying webbing around a tree and clipping the midpoint of the 200-foot rope through a locking carabiner. We then throw the two ends of the rope over the cliff. This way, when we rap down to a suitable anchoring point between us and the victim, we can pull the rope down and do it again. The downside is that using the doubled-over 200-footer, we can rappel only a hundred feet.

As I start over the edge, I flap the ends of the rope to see if they reach a reasonable anchoring spot. I can't tell. I climb back up to the anchor and tell Casey I'm not sure the rope reaches a good landing. Clipping in, he goes to the edge and looks down. He's pretty sure they reach. I know I'll really be screwed if I rappel down a hundred feet and find myself still twenty feet off the deck! I also know that Casey, with twice as many eyes, has better depth perception than I, so I rap down—sure enough, the ropes make it to the next shelf, with twenty feet to spare. Casey raps down and pulls the rope down after him.

We are still about seventy-five feet above the victim. The only anchor is a lone (although bomber) tree fifty feet above us on another rock outcropping. We climb a snow ramp up to the tree, setup another anchor, and again clip the midpoint of the rope into the anchor. I rappel down about hundred feet to the victim.

I've been giving periodic updates to the incident commander throughout our approach. When I reach the victim, I immediately say, "Command, this is Steve. I've arrived at the patient. Stand by for a medical report."

After introducing myself, I begin my medical assessment by asking him what happened. He mumbles that he's cold. I take his vital signs and ask him the SAMPLE history questions, then do a rapid head-to-toe exam.

After completing my assessment, I call the command post and surprise them with my summarized medical report: "The patient is alert and oriented times four. He is suffering from mild to moderate hypothermia."

I can hear the silence on the radio. We were about to give this guy up for dead, and here I am, telling them he's basically okay. Of course, "basically" is a relative term—he's obviously very cold and exhausted.

In emergency medical care parlance, "mild hypothermia" means the patient is still able to shiver and that his body temperature is above ninety degrees Fahrenheit. Although I don't have a thermometer, this guy is thinking pretty well and still shivering. The risk is that as a hypothermic patient gets colder and runs out of energy, he will become too exhausted to shiver; then his body temperature will drop rapidly.

I look at him and think, *you are one tough cookie!* I don't think the average person could have survived what this man has and still be conscious—I don't think I could.

I am wearing long johns, fleece insulation, Gore-Tex pants and jacket, and mountaineering gloves, and I'm feeling the chill. This guy is in a T-shirt, running shorts, and tennis shoes, his clothes are soaked from spending the night in a snowstorm, and he's been sitting still!

The first thing I do is make an improvised climbing harness for my guest. I want to get him tied in and secure as soon as possible. I do this by tying two leg-size loops in a twenty-foot length of webbing and having him stick his legs through the loops. He can't quite lift his legs, so getting them through the loops is a little like trying to dress a toddler in pants. He is exhausted and passive. I wrap the webbing around his waist several times and secure the ends in front with a water knot. As soon as he has a harness on, I clip a pigtail from my harness to his. Although his improvised harness lacks the padding and comfort of a commercial rig, at least he isn't going to fall off the cliff.

It has taken Casey and me several hours to reach the patient, and we brought only a minimal amount of gear. On most rescues, there's a stream of rescuers bringing in additional gear, but this time, we'll have only the gear Casey and I carried while climbing.

To stabilize this patient medically, I really need to get him into warm clothes. But we don't have any extra clothes, and what we do have is wet from snow, sleet, and sweat. I take off my fleece jacket and wrap it around him—that will have to do until we can get him down off this rock wall.

Casey rappels down the rope to join us. He ties into the ends of the rope, and rigs our second rope so he can lower us its full length: 180 feet. I connect the patient and myself to the second rope and use a Prusik to position him a foot above me.

I can see two Utah County rescuers below us. One is C. J., who is pacing back and forth. I call him on the radio and ask him to estimate the distance from the edge, which is still twenty to thirty feet below me, to the ground. He tells me it is 100 to 130 feet. Granted, these distances are only estimates, but if the edge is 30 feet below me and it is an additional 130 feet to the ground, we will need 160 feet of rope. That leaves a margin of error of only twenty feet. I'm hoping C. J. is good at estimating distances.

Almost all urban rope rescue teams will use two ropes during a vertical rope rescue: one rope for lowering and the other as a backup belay should the first rope fail. A few mountain rescue teams use a single rope during vertical rescues. Sometimes the decision to use just one rope is influenced by some combination of terrain, equipment availability, confidence in the rope and anchor system, and time constraints such as approaching weather or a patient's deteriorating condition. Unfortunately, Casey and I don't have the luxury of two separate ropes—we barely have enough rope to reach the ground.

Our heavy-duty rescue rope is now connected to a tree above us and doubled over. It can hold 6,000 pounds per strand, so 12,000 pounds doubled. That rope got us to our current location. Casey and our second rope are connected to the ends of that rope. The second rope is a much thinner, stretchier climbing rope. This rope is called a "half rope," not because it is half the normal length but because it is not intended to support a falling climber with just one strand—you're supposed to use two of these specialized half ropes when climbing.

I state the obvious to Casey: that we'll be using single-rope technique with a skinny rope not built for the task. Casey grins and reminds me, "That climbing rope is *so* strong." I agree, the skinny climbing rope is amply strong to hold two people, and we really want to get our patient off this wall.

Casey and I confirm that my belay is on. The patient is too exhausted to stand up, but at least he can squat, so I put my arms under his armpits and gently drag him to the edge. He tries to help us along by pushing with his feet.

I choose to have Casey lower the two of us, rather than tie the victim to me and have me rappel our combined weight. This way, I will have both hands free to work with the rock and position the patient. I may also need to provide medical care. And I'll want Casey to have control of the ropes so the patient can't interfere with our descent. If the patient grabs the rope or punches me in the nose—not out of the question for a hypothermically compromised patient—our descent will still be controlled by Casey.

At this point, Casey is holding less than half our weight, because much of it is still on the steepening ground. As I get nearer the edge, I realize that the transition from half-walking and half-hanging to completely hanging on the rope, will be difficult. The transition from standing on your feet to sitting in your climbing harness, with your weight on the rope, is almost always the hardest part of rappelling. This will be considerably harder, because there are two of us tied together, and one is barely ambulatory.

As we near the edge, I see that the rocks have many sharp edges. Ropes are extremely strong, but under tension they can be cut with alarming ease. I call up to Casey, "Stop!" Laying the patient on the ground at my feet, where he half hangs on the rope, I pull my heavy-duty space blanket out of my backpack and lay it over the sharp rocks to protect the rope. Then I struggle to lift my patient. "Down slow!" I tell Casey.

Much of the patient's weight is now being held by the Prusik that connects him to the rope. I pull him into my arms and press my feet against the steepening wall, trying to keep us both off the rock.

I lean back, and we sort of roll over the edge. It's an unnerving feeling, dangling from the skinny rope, as he slides over the edge and pulls me over. The falling-helplessly-over-the-edge feeling reminds me so much of jumping out the door of an airplane that I actually think of parachuting as we "fall" over into space.

Once we are on the vertical wall, I can position myself slightly below the patient and cradle him in my arms like a baby, with his weight held by the rope above me.

Casey slowly lowers us another twenty or thirty feet. I can see another edge coming up that looks to be at least a hundred feet above the ground, there's C. J., still pacing furiously below. This edge has rocks with some wicked-looking edges.

I now have two concerns. One is rockfall. If a rock happens to fall and hit my rope where it crosses this edge, it will almost certainly cut the rope. Yes, the likelihood that a rock will land on the rope is small, but there's a lot of loose rock, and the possibility is real.

My second and much greater concern is that the thin climbing rope, now being pulled guitar-string tight by almost four hundred pounds of rescuer and victim, might be cut by the sharp rocks on this second edge.

As we are hanging on the rope, I recall images of a demonstration that is performed at most rope rescue classes, in which the instructor ties one end of a rope around a tree and has three or four rescuers pull against it. This puts a few hundred pounds of tension on the rope. Then the instructor takes a pocketknife and touches the edge against the rope—the rope snaps like a thread.

I call to Casey to stop. We're about ten feet above the second edge and still more than a hundred off the ground. I think I can probably go over the sharp edge without cutting the rope. *Probably.*

I consider the likelihood of the rope being cut, and the resulting consequences. Also complicating my situation is the narrow range of options. In fact, I can only come up with one solution, which includes risks of its own.

There is a risk metaphor using Swiss cheese. If the cheese represents safety, it's okay if there are a lot of holes in the cheese. The real risk comes if the holes in the cheese line up, and the more holes in the cheese, the more likely they will.

It's one thing to be on a rock face covered with snow, another to be tied to a lethargic victim suffering from hypothermia, another to be hanging on a rope a hundred feet above the ground, another to have that rope be a skinny "half rope," and yet another to encounter a second set of sharp rock edges with no way to protect the rope. There are a lot of holes in this cheese. I can accept a few of these risk factors, but I'm not sure I can accept all of them at once.

A strange scenario crosses my mind's eye. In it, I am dead, and the team is having a debriefing on the mission, trying to understand what happened. I can hear my good friend and teammate Tom Moyer saying, "I can't believe Steve was dumb enough to hang two people on a single nine-millimeter rope over a sharp rock edge. What was he thinking?" And he'd say it just like that, too. And I would roll over in my body bag from embarrassment.

That thought is enough for me! Although I might die while preventing the rope from being cut, I don't want people to say I wasn't thinking.

There's an opening in the rock next to me, about three feet wide and two feet deep. I manage to push the patient into this crack, jam my right foot against the far wall, put my right knee under his bum, and put my left foot on the wall behind me, then sit on that foot. This wedges us into the rock, with the patient leaning into the crack and sitting on my knee.

I explain to him, "I need you to stay perfectly still for fifteen minutes. Can you do that for me?" In his numb state, he mumbles, "Yes." I speak again, more forcefully, "We are going to have to come off the rope. You *cannot move at all.* Not at all. Do you understand?" This time he's more alert. "Yeah, I understand," he says. "You guys do this all the time. I'm okay with it." Now, I appreciate the vote of confidence, and I have done my share of rope rescues, but I most certainly do not do *this* all the time!

More holes in the Swiss cheese; we are tired, my patient is very cold, the rock is loose, and I am coming off the rope.

I call up to Casey on my radio. He doesn't answer. I try again. No answer. "Casey!" I yell, "Turn on your radio!" I hear his reply from above, "My battery's dead." Another hole in the cheese.

I lean my head back and yell to Casey about my predicament, "I'm concerned about the nine and the sharp rocks. I want you to reset the system so the nine is doubled on the tree and you are lowering me on the eleven."

The "eleven" is the eleven-millimeter rescue rope, which is not only much stronger than the nine-millimeter climbing rope, it's also much more resistant to cutting and abrasion.

I can barely see Casey as he strains his neck to look over the rocky edge above me. His mouth says, "What?" but his tone says, "Are you frigging crazy, Steve?"

"I'm serious, Casey. This isn't safe."

"Okay." I can hear that he is resolved to doing this even though my plan seems nuts to him. Of course, he doesn't see the knife-edged rock below me.

I untie the victim and myself from the end of the rope and hold the end out for Casey. "Up rope!" The tail of the rope moves jerkily up the rock wall.

C. J.'s pacing down below seems to have doubled in speed. He looks up at us, says something to his companion, looks at the ground, and continues pacing.

After pulling up the rope, Casey climbs to the upper anchor. He finds another tree that, although much smaller, is a little lower on the ridge. This will buy us a little more length, but Casey wisely chooses to postpone telling me about the smaller tree until a week later. He'll say something like "I didn't want to freak you out over the little tree, but we were both unsure if the rope would reach the ground." He connects the smaller climbing rope to this tree.

He rappels down the now-doubled climbing rope and connects himself, and the rescue rope, to its ends. He then throws one end of the rescue rope over the edge. The end doesn't quite make it to me, but I can just reach a loop and pull the tail down, keeping my head tucked into the crack to avoid being hit by the small rocks I'm dislodging.

After tying into its end, I call, "Up rope!" to Casey. I can barely hear his reply, but I see the rope move jerkily upward. When it is taut, I call, "On belay?" to Casey. He replies, "Belay on!" Those are beautiful words.

I push my legs against the wall to extract my partner and me from our temporary perch. We immediately start moving down the rock face as the ropes stretch under our load.

I hear Casey yelling, "Stop! Stop! Stop!"

It turns out that as I shifted our weight onto the ropes, the upper rope, which is now a stretchy climbing rope, began elongating and slowly pulled Casey toward the edge. He was still safe, since he can go over the edge and still lower us while dangling on the rope, but I'm sure the feeling of being unexpectedly pulled over the edge must have been a little unnerving.

After Casey ties in farther up the nine, I again shift our weight onto the rope. I then give Casey the "Down slow!" command, and he lowers us over the edge. This remains a little exciting, since we are still high above the ground, but it feels so much better to be on this durable rescue rope.

Once over the second edge and with the victim hanging above me, I again turn him so I can cradle him like a baby and use my feet to keep us off the wall. I do a silent prayer to Mother Earth, asking that she not cut our rope and receive us too abruptly. I don't like to be in situations that inspire me to pray, but this seems an appropriate time.

The final ten feet of this wall are overhung, like a cave. I keep my feet on the upper edge of the "cave" so we don't bump into the wall. When we are below the overhang, I pull my feet in and we swing gently under the roof. There, in this cavelike hollow, are my teammates, Keith and Jared. I am stoked.

It is great seeing them, as well as numerous rescuers from Utah County. It feels good to know that we've completed our part of a challenging mission and that people I trust are taking over.

Many hands reach up and gently lower us to the ground. Rigged as we are for the vertical environment, we are not graceful on the ground, with our harnesses tied together, lying there like dead marlin. I am now treated more like a patient than a rescuer, as multiple rescuers begin untying the knots and unclipping carabiners. They even help get me out of my backpack.

The additional rescuers then help the victim down a moderately steep 150-foot scree slope. They do this by connecting a short rope between a rescuer and the victim, and then the two of them are slowly lowered as they walk down the slope. They choose to walk him rather than put him in a stretcher, because they are concerned that putting him in a stretcher might cause his body to cool down even more. He is just not generating much heat on his own, and they want him to get a little exercise, while on rope and supported by a rescuer, to warm him up. I'm stunned to see that he can hold himself up with just a little support. This guy is *tough.*

When they get to the bottom of the scree slope, they load him into a litter with a single wheel, and additional rescuers wheel him down the trail to a waiting ambulance.

As they are doing this, I walk from the cavelike overhang over to C. J. He looks at me and wags his head. "You've got bigger balls than me," he says. I take it as a compliment, but the truth is, mine were feeling pretty small up there. C. J.'s teammate comes over and tells me, "C. J. kept pacing back and forth and mumbling, 'I don't have a good feeling about this. I don't have a good feeling about this.'" I can understand why. We are in C. J.'s county and he's the team leader. He is responsible for everybody's safety on this mission. Here he is, with two guys he barely knows, on his watch, executing a potentially risky rescue on snow-covered cliffs.

While I'm talking to C. J. and the patient is being lowered to the valley floor, Casey calls down several times, saying he is cold. I call up to him and ask him to wait until the victim is no longer below him. We don't want him dislodging rocks onto our almost-safe patient. I can understand why he is cold, and expect that he may well be getting hypothermic. The weather is lousy, and he hasn't been moving very much. I have at least been wrestling the patient and also got a hefty dose of adrenaline to warm me.

Hypothermia is a real concern. The Utah County rescuer who was transported to the hospital this morning was found in the meadow, completely disoriented, with no idea where she was or what was happening. We certainly don't want Casey managing ropes and rappelling down a cliff in that condition. The oft-repeated rescuer's motto is, "Don't become a victim." To adhere to this, most rescue teams put their safety and the safety of the victims in what may seem an unusual order of importance. The protocol says, "*Your* safety comes first, then the safety of your teammates, and then the safety of the victim." At first blush, this may sound selfish and uncaring. But the fact is, the best way to look after your patient is to have a healthy rescue team, and the best way to keep the team healthy is to be ever mindful of your own safety.

When Casey reaches the ground, I walk up to him and give him a big handshake. We are wearing huge smiles. We look up at the rock face with the single, tan rope running down its face. The ropes will have to stay. As we are looking up, my silver and red space blanket comes drifting down like a cut-away parachute. As it floats down, I feel an eerie sensation of falling. I'm glad we are safely on the ground.

It's hard to believe that we almost called off the rescue and left this man for dead. Given the information at the time, calling off the search may have been the right call, but it was the stubborn perseverance of the Utah County rescuers, who kept calling for the victim until the very end. They made the difference between a save and just another sad statistic.

I'm not proud that I needed to have Casey rerig the system in mid rescue. I certainly wish I had changed it before going over the first edge. Of course, Casey and I had discussed using the single climbing rope, and we both felt that it was appropriate—we wanted to get our patient off the mountain quickly. We just didn't factor in the sharp rock edges. Although I'm a little chagrined that I had to stop the mission to rerig, I am glad that I was willing to stop, admit that I screwed up, and get it right. I'm even happier that it worked.

I figure each rescuer can do a few of these risky rescues during his or her lifetime, but not many. Just as a cat has nine lives, we have to look at the few high-risk rescues that we each have as a finite (though unknown) number and be judicious in doling them out. Now, you might choose to do one of these higher-risk rescues when you are trying to save your own climbing partner, but you *cannot* take these risks repeatedly as a professional rescuer—eventually, the odds will catch up with you. The media loves to show rescuers and firefighters as "risking their lives to save others." And rescuers and firefighters often do things that look risky, but the truth is, professionals control the risks such that, although it may look bold and daring, their lives are not at significant risk—well, rarely, anyway.

The exposed cliffs traversed by rescuers
(Visit MountainResponder.com/Photos and enter Story Code UCC.)

Eleven
Risking

The reasons people need to be rescued can be loosely grouped into three categories. They made big mistakes, smaller mistakes, or no mistakes and simply fell victim to random misfortune. Of course, most rescues result from a *combination* of mistakes and misfortune, but they broadly fall into these three categories. The result is the same; someone calls 911.

Big Mistakes

It's usually obvious when a victim makes a big mistake. I'm talking about someone who does something so foolish, so fundamentally reckless, that even without the benefit of hindsight, almost everybody would have asked, "What are you *thinking?*"

Some big mistakes appear to be the result of ignorance—the victims simply don't have a clue what they are getting into. For example, people who go backcountry skiing or snowshoeing when there is there is significant avalanche danger, without avalanche safety gear or training, may have no idea that they are making a big mistake. In fact, five of the nine avalanche fatalities I've responded to involved people who were not wearing an avalanche transceiver. That corresponds to the national average, which shows that about two-thirds of avalanche victims weren't wearing a transceiver. The transceiver might not have saved them, but entering avalanche terrain without the basic tools is a big mistake.

It's hard to know whether people who climb on near-vertical rock without the appropriate climbing gear and training are ignorant or just feeling lucky. When I arrive at a call like the one for someone geckoed to a rock wall, and someone else seriously injured on the ground, it sure looks like a big mistake.

And sometimes it's someone else's mistake, big or small, that results in injuries to others.

```
Callout S & R for multiple injured
persons in Mineral Fork 7550 e Little
Cottonwood Canyon. Staging point is
of Mineral Fork. Come 1041 on Tac One.
13:53 08/19/06
```

When the rescuers arrived on scene, they discovered that three children, from six to ten years old, and their dad had been riding on one ATV. I think most people would say that putting four people on one ATV is, in itself, a big mistake. When Dad then accidently drove the ATV off a twelve-foot ledge, it probably became clear to him that it wasn't so smart. Dad had a possible fractured clavicle and a concussion, which was probably the result of his not wearing a helmet—another mistake. The kids got a lot of bumps and bruises but were released by the paramedics, while Dad was airlifted to the hospital.

Many, perhaps most, of the big mistakes I saw had contributing factors. Consider this callout for a "lost hiker" up Neff's Canyon:

```
SEARCH & RESCUE CALL OUT.LOST HIKERS,
NEFFS CYN,4100 S 4900 E. CONT CAR 600
ON CODE 1,WINTER HIKING GEAR NEEDED,
CS 5151,EOM
18:38 01/15/04
```

As I was driving to the trailhead, a radio request said, "We need a hasty team ASAP. The complainant says his friend is dying." When I arrived, I quickly stuffed my backpack and took off with my partner. The snow was firm, so we went in wearing hiking boots rather than skis. Due to the reported urgency, we alternated between a fast hike and a slow jog.

The sketchy reports we received said the missing person went into the mountains at one p.m. and had been drinking. We were also given a description of the sole of his boot, and, like trackers but without their expertise, we would stop every few minutes, locate the footprint, and resume our jog.

Each time we stopped, we also called out the victim's name, "Bobby!" After a few miles, we heard his reply: "Hey, you motherfuckers!"

I gave the update over my radio: "Command post, this is Nine-oh-one. We have voice contact. He sounds belligerent." My partner and I agreed that for our own safety, we wouldn't mention that we were with the sheriff's office, just that we were a rescue team from the county.

The "lost" man was sitting in the snow on a hill above us. I talked to him, interspersing my conversation with four-letter words to gain his respect. He explained that he was fired earlier in the day. My partner and I did our best to show some compassion, but it was a little difficult, considering he had held his dishwashing job for only six months.

The cold was taking its toll on him. Ice was stuck to his ungloved hands and his Levi's jeans were frozen into rigid denim tubes. We couldn't tell how much of his slurred voice was the result of hypothermia and how much was from drinking.

Yeah, this guy made a big mistake. He got drunk, hiked into the snowy mountains in street clothes, and got lost and hypothermic. It wouldn't take tremendous wisdom to avoid his situation.

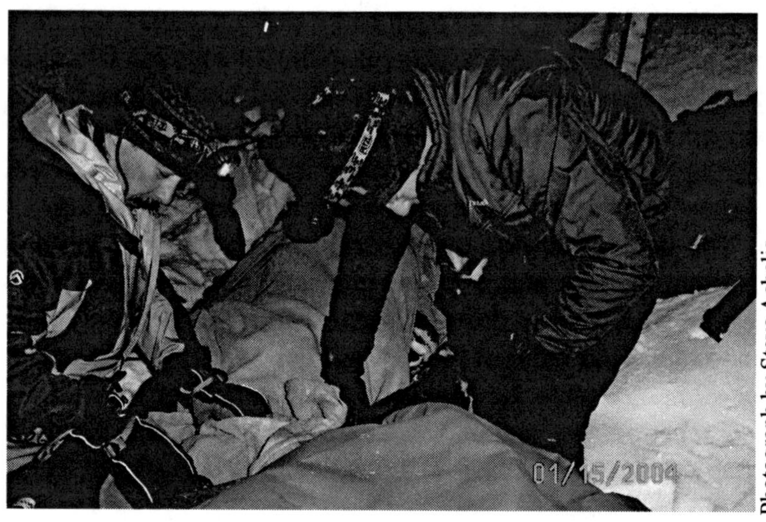

Rescuers package the hypothermic patient
(Visit MountainResponder.com/Photos and enter Story Code NCD.)

Drugs and alcohol were contributing factors in many of the big mistakes I witnessed. There was the young man who put down his six-pack of beer to cross a raging creek. As he was swept over a cascading waterfall, a submerged log caught his pants and yanked them down to his ankles, and there he drowned, hanging upside down in a waterfall. And there was the young man found dead at the base of a rock wall, with a bag of weed in his pocket. Big mistakes made under the influence of mind-altering substances.

Smaller Mistakes

When I talk to people who don't spend much time in the mountains, they tend to think most people get into trouble because they make big, dumb mistakes. "Why do you help those yahoos?" they sometimes ask. But in my experience, it's often people making smaller mistakes—or, more commonly, a *series* of smaller mistakes—who get into trouble.

Many of these mistakes may not have looked like poor decisions at the time, but in hindsight, it becomes clear that a small decision that seemed inconsequential at the time led to their predicament. For example, somebody goes on a hike, doesn't bring enough water, and gets severely dehydrated. Going up without enough water probably didn't seem like a big deal when they left the trailhead, but it sure looks like a mistake when they are lying in the dirt in misery.

Sometimes someone makes one small mistake—a mistake so small it might be considered simple misfortune—with real consequences.

```
S&R CALLOUT AT BRIGHTON P/L UP BIG
COTTONWOOD CANYON ON A FALLEN HIKER ON
WASATCH CREST GUARDSMAN PASS CS 77861
COME 10-8 ON CD1 EOM
13:48 07/04/02
```

When we arrived on the mountain pass, we found a mountain biker. His front wheel had come off and his front fork dug into the soil, flipping him over the handlebars to land on his head. Even though he wore a helmet, his jaw and right orbit appeared to be fractured. We carried the patient to a nearby saddle, where an Airmed helicopter could land and fly him to the hospital. Some might call the accident a random misfortune, but perhaps it was a small mistake not to check the quick release on his front wheel.

Small mistakes have a way of compounding, like those risk-holes in the Swiss cheese. And experienced backcountry recreationalists aren't exempt. I was winter camping by myself when a storm came in twelve hours early. After wasting my headlamp batteries reading a terrifying story by Dean Koontz, I broke camp at one a.m. to escape the increasing avalanche danger. While skiing out, my headlamp died and I mistakenly followed ski tracks to the edge of a cliff. The result of compounding my otherwise inconsequential small mistakes? I traversed significant avalanche danger, ending up bivied under a tree, and became hypothermic. Fortunately, I was able ski out and warm myself in my truck.

No Mistakes

Finally, there are people who, through no fault of their own, have something bad happen. Misfortune. The classic example is a backcountry skier who falls and injures her knee. I guess some people will say that skiing is risky, but so is driving a car. The only way to avoid all ski injuries would be to not ski.

Rockfall at Donut Falls

In May 2003, I get a page that sounds too strange to believe. A hiker has been pinned by a rock but is "still breathing." I hadn't heard of anyone being trapped by a rock until I read about Aron Ralston, who got national attention after being pinned by a boulder and cutting off his own arm. And this call comes just three weeks after Aron's ordeal. I am puzzled.

The call says the accident happened at Donut Falls. This waterfall is a popular destination just three-quarters of a mile from the parking lot. It's the kind of hike you can take your granny and your toddler on.

I drive fast, breaking the rule against exceeding the speed limit when responding to callouts. Sorry, boss, but the "still breathing" report carried the strong implication that he might not continue doing so if somebody doesn't come to his aid pronto.

Driving up the winding canyon road, I get stuck behind a fire truck. It's one of those long aerial-ladder jobs, with a huge telescoping ladder on top and a "bucket" on the end, where a firefighter can stand to spray water. They're going really slow, probably twenty-five in a forty zone, with their lights flashing and sirens blaring. I pass them on a straight stretch of road and hold my badge out my window for them to see, and they still lay into the horn and flap their arms as if to say, "You can't pass an emergency vehicle!" *Good grief, mates, I'm trying to help the same guy you are. And if he's at Donut Falls, I don't think he needs a ladder truck.*

After reaching the trailhead, I drive up the narrow dirt access road. Brighton Ski Patrol Director Patrick Eibs, moonlighting as a firefighter, is using a chain saw to clear downed logs from the road.

It's May, and large snow patches still cover much of the road. Even in four-wheel drive, I'm slipping all over. I park the truck and throw my gear into Darren Westerfield's Suburban, which is able to claw its way up the road.

After parking and putting on our backpacks, we hike the short trail to the waterfall. The water pours through a hole in the rock: hence the name, Donut Falls. Occasionally we have "donut divers," who climb above the falls, slip, and zip right down through the hole. Big mistake or little mistake, it's one with dire consequences.

I hike up a small snowfield to the left of the creek and cross a snow ramp. Although I have to cross only fifty feet of snow, it would be safer with crampons and an ice ax. If I should slip, how would my teammates judge the size of that mistake? Anyway, on the other side of the snowfield I find the young man, with an enormous boulder, probably four feet in diameter, on his back. He is folded in half, with his head between his knees and the rock on his back. His feet are in the reflecting pool.

It looks as though he was sitting down when the boulder landed on him. Here is a summary news report from the *Salt Lake Tribune*:

> *In the last moments of his life, a 22-year-old Salt Lake City college student saved a friend from harm when he pushed her out of the way of a falling 5-ton boulder. Unable to escape himself, Seth Buhr was hit and killed by the rock.*

Officials of the Salt Lake County Sheriff's Office said four friends were sitting and talking by a pool at Donut Falls in Big Cottonwood Canyon when a boulder above them came loose at around 1:30 p.m. Buhr saw the tumbling rock and pushed his female friend into the pool of water, out of the boulder's path. He was crushed by the rock.

This young man not only didn't make any mistakes, he also saved someone else's life during the last seconds of his. Impressive.

Rescuers remove the multi-ton boulder
(Visit MountainResponder.com/Photos and enter Story Code DFR.)

Likelihood and Consequences

Risk taking is a combination of considering the *likelihood* of something going wrong, and the *consequences* if it should. You can make an obvious mistake, like going on a ten-hour hike without any food, yet have minor consequences—you just get really hungry. Or you can do something that should be quite safe, like hiking near a picnic area, yet slip and end up with significant consequences—a broken leg.

Aron Ralston's self-rescue, in which he amputated his arm with a multitool, is certainly impressive. And the likelihood of something going wrong on his trip was relatively low. But the consequences of his not telling anyone where he was going were that no one came looking for him and he was forced to cut off his arm.

A twenty-one-year-old woman hiked up Ferguson Canyon to swing on a rope swing. After watching her male friend use the swing, she probably never considered the likelihood that she would be unable to hold on to the rope—or the consequences if she didn't. When we arrived, she was in a near-vegetative state, from which she would never return, after falling and hitting her head on a rock.

When Casey was lowering the hypothermic patient and me down the vertical wall, the likelihood of the thin climbing rope being cut on the sharp rock was relatively low, but the consequences would have been two fatalities.

The Right to Take Risks

One might expect me to take a cynical view of people making choices that result in their needing rescue. Sometimes I did. Yet I also became more accepting of people's right to take risks. The result of our society doing such a great job of removing risks from our daily lives, from mandating seat belts to bicycle helmets, makes it understandable that some people may want to put themselves in risky situations in the mountains. They have a desire—a need, even—to bathe in the heady, healthy feeling of freedom and of getting closer to life's edge, and to get a break from the monotony of day-to-day living. It probably goes without saying, but most rescuers wouldn't be in the business if they themselves didn't enjoy pushing the envelope.

Personal Risk Acceptance

But along with the right to take risks comes the responsibility to plan on self-rescue. No one should go into the backcountry expecting the government to save them should they run into trouble. Maybe in the city, but not in the mountains.

The number of rescues in Salt Lake County is not increasing, even though more people than ever before are in the backcountry and more are carrying cell phones. This implies that people are managing the risks and doing more self-rescue. Of course, if you need help, you should certainly call 911. Planning for self-rescue doesn't mean you should macho it out and not ask for help when you need it.

After evaluating the likelihood of something going wrong, and the consequences if it does, we get to decide if we want to participate in an activity "out there." My own willingness to accept risk has been tempered by the blood, broken bones, and body bags I've seen. That doesn't mean I don't do things that some might consider risky, but I do think more about the risks and take steps to minimize them.

For example, when I grab a handhold while climbing, I am more careful than I once was to make sure the hold is solid before I put weight on it—easy to remember after seeing a climber take a bad fall with the hold still in his hand. When I communicate with my belayer, I make sure we hear each other, because I've seen poor communication result in too many broken bones. And when I'm planning to go backcountry skiing, I pay close attention to the avalanche report and snow conditions.

We received a call for two young women who were stranded in the mountains.

```
This is a callout for stranded
hikers..7900 e Millcreek cyn...Elbow
Fork..respond on Code One
Frequency...case#102108..eom
17:17 08/30/03
```

It took us a few hours to hike to their location. The women were sitting down on a slope of large talus. We walked up to them, asked if they were okay, and I began asking the basic SAMPLE questions to make sure this wasn't a medical call. Finally, one of them blurted out, "Could you please not stand there? You're scaring us!" It was easy terrain for us to walk on. But to them, the fear of falling was petrifying. This was a good reminder that we each have our own limits and that those limits merit respect. Stopping when you reach your limits is a good thing.

We attached ropes to the women—for their confidence, not for the conditions—and walked them down the rocky terrain to the trail.

Disbelief

Far and away the most common sentiment I have heard expressed by victims, or friends and family of victims, is that "I can't believe this is happening to me (or to us)." They repeat this over and over, like a mantra. When their climbing protection fails, when their friend disappears into the whitewater, or when we are probing the snow for their ski partner, it's the perennial refrain: "I can't believe this is happening." I think that, consciously or subconsciously, they actually *have* considered the possibility of an accident and decided that it will be okay. And then, when things turn out otherwise, they just can't accept that they were mistaken and that the consequences are so severe.

Twelve
Stable or Dead

In backcountry search and rescue, there is a cliché that our patients are either stable or dead. That's because it takes so much time to locate and access the patient in remote terrain, they must be stable, or they'll die before we reach them. I have had only a handful of patients who were neither stable nor dead. This chapter tells about two of those patients.

It Takes Heroes

July and August can be scorching hot in Salt Lake County. Today is July 29, and we've already had nine days when the temperature topped a hundred degrees. Today it is ninety-eight.

Jim Guilkey and Carol Butler were hiking above Lake Blanche in Big Cottonwood Canyon. The Lake Blanche trail is three miles long and climbs 2,700 feet. They were up-canyon from the lake when they heard rockfall. A little earlier in the day they had met a man on the trail. I'm not sure if they heard a cry for help or just got curious about what might have prompted the rockfall, but they hiked toward the sound.

There they found a man—a Russian, though he spoke excellent English—who had fallen. It wasn't a huge fall, as falls on rock go. He had tumbled only about twenty feet, but on sharp, angular rocks, most of which were the size of a personal computer or larger. This was the man they had met earlier that day, now injured and unable to hike down.

They couldn't get cell phone service there, so Jim hiked up the ridge while Carol stayed with the injured man. After he expressed concern about getting sunburned, Carol tucked paper napkins into the bottoms of his pant legs, to keep the sun off his exposed ankles. He sounded a little delirious, repeatedly bringing up his concern about getting sunburned. And Carol, no doubt feeling helpless, accommodated his requests.

He also had a painful cut on his hip, so Carol tucked a paper napkin under his waistband to protect the wound. Meanwhile, on the ridge above, Jim was able to get cell service. He called 911.

Tom Moyer and I are sitting at the top of a spectacular four-pitch rock climb, taking in the view, when both our pagers go off. The page displays the notoriously ambiguous words "fallen hiker." All we really know is that we don't know what to expect. We rappel down from the ridgetop, hike to my truck, and drive to the nearby staging area.

At the trailhead, we find several deputies and a fire engine. One of the deputies is talking to Jim on his cell phone and getting preliminary information.

It takes the average hiker two or three hours to hike to Lake Blanche; carrying our rescue gear, it would take us longer still. We decide to use a helicopter both to locate and to access the victim.

We first call Life Flight to see if the hoist ship is available. We would prefer their more powerful helicopter and would like the option of having the hoist. Unfortunately, their hoist-equipped helicopter is on another call.

The deputy relays the patient information he received on the cell phone. The victim has a cut on his hip and may have a head injury. Jim tells him there's a grassy area within walking distance of the victim, where he thinks a helicopter can land.

Tom and I are the first rescuers to arrive. Because we will be the hasty team, I pack medical gear to stabilize our prospective patient, and Tom packs basic rope rescue gear to transport him. Additional teams will bring in more gear once we have a better understanding of the mission. We also bring enough personal gear so that we can hike out on our own.

Any rescuer who is inserted by helicopter must have the skills and gear to get out of the backcountry alone, and to spend the night if necessary. This is a basic rule of rescue that must be understood before climbing into a helicopter—this ride is guaranteed for one way only.

Deputies shuttle us up the winding canyon road in their SUVs, to a location where the helicopter can land; then they close the road by parking their vehicles above and below the LZ. Tom and I hunker down on the side of the road with our backpacks while the brown DPS helicopter lands on the narrow two-lane road. We stay put till the pilot makes eye contact and nods his head. Despite the pilot's sunglasses and earphones, I recognize him at once—it's Steve Rugg, whom I've done several missions with.

We approach the helicopter from the side. It's standard safety policy that you never approach a helicopter from the rear, because the pilot can't see you and because the tail rotor spins so fast, it becomes almost invisible. Stories, not all of them mythical, tell of unwary passengers suddenly dissolving into pink mist as they walked into an unseen tail rotor and were pureed from the waist up. Stooped over, we half drag, half carry our backpacks toward the ship.

When we are twenty feet away, I see Steve shaking his head and holding up one finger. I get the message—he can take only one of us. I turn around and give Tom the same shake of the head, a bit of a frown, and hold up one finger. He smiles, turns around, and drags his pack away from the helicopter.

It doesn't surprise me that Steve can take only one of us. Just last week we had two hikers who slipped on a snowfield high on a nearby mountain, and because of the heat and elevation, the Life Flight helicopter was unable to insert even one rescuer.

I open the door, put my backpack on the rear floor, and climb into the front seat next to the pilot. Once I put on the headphones, Steve apologizes for having to leave Tom behind. No apology needed—he's saving me a hike of three miles and more than three thousand vertical feet. And he'll be bringing Tom in shortly.

Making a pass above the lake, we locate the three people—the fallen hiker and two bystanders—on a steep talus field. There's a patch of grass below them just large enough to land the helicopter. Steve sets down and asks if I want him to stay. Jim had told the deputy he thought the helicopter could land "within walking distance." I'm hoping that means we can walk the patient to the helicopter. After all, it sounds as though his only injury may be the cut on his hip. Over the comm line, I ask Steve to wait.

I climb out of the helicopter, pull out my backpack, and drag it across the grass. Once I'm clear of the rotors, I strap it on and start hiking up the loose rock toward the trio. Jim walks down to meet me.

I introduce myself and ask Jim if the injured person can walk down to the helicopter. He shakes his head adamantly and says, "No way." I feel a little dumb for suggesting it, but the injuries didn't sound too severe. I key the microphone and tell Steve that I will need additional rescuers.

It takes only a few minutes to reach the fallen thirty-two-year-old hiker, lying awkwardly on the rocky slope. Dried blood, darkened to almost black, covers half his face and is matted in his hair. His right eye appears to be glued shut with blood.

I introduce myself: "Hi, my name is Steve and I'm with the Salt Lake County Sheriff's Search and Rescue Team. What happened?"

He manages to open his eyes as I ask this question, but they quickly close. I'm not yet sure whether his eyes closed because of a lowered level of responsiveness or because of the bright sun. All he can tell me about the accident is, "I fell."

Complications

While talking to him, I pull on a pair of blue medical gloves and start my rapid trauma assessment. With eyes and hands, I will quickly assess his body, looking for significant injuries. The head-to-toe exam should take less than two minutes.

While checking his head, I find several scalp and face lacerations, which have covered his face and hair with blood. Scalp wounds can bleed profusely, but the active bleeding has stopped. While inspecting the outside of his head, I look for signs of a serious head injury. As I do, I recall feeling a loose section of skull when I palpated a fallen climber. I also remember a fallen climber who had clear cerebrospinal fluid leaking from his ear. Happily, I find neither of these ominous signs with this patient. Without obvious signs of a serious head injury, all I can do is monitor him to see if increasing pressure within the skull alters his level of responsiveness.

I quickly check his eyes, mouth, ears, chest, and abdomen as I work my way down his body. Jim and Carol are telling me about how they heard the rockfall, came over, and found the guy, but I'm tuning them out.

When I reach his waist, I see a white paper napkin sticking out from his pants, above his right buttock. I ask Carol what the paper is, and she tells me he has a cut on his hip. She put the napkin there where she saw blood. I don't see any blood on the napkin and almost continue my assessment down his legs, but in my mind I can hear DeAnn Barnson, one of my EMT instructors, admonishing me to "expose every injury." I don't really want to—after all, it's just a little cut, and to expose it I'll need to undo his belt, unzip his pants, and find a way to slide his pants down without moving his spine.

Modesty certainly should not come before patient care, but I also know that to get a clear view of his buttocks, I will need to pull down (or cut off) his pants, which will expose his genitals to the sun, Jim, and Carol.

Now, urban EMS providers will pull out their shears at the drop of a hat and cut away clothing to expose injuries. Mountain rescuers are much slower to do so, since clothes provide important protection from the elements.

Okay, DeAnn, I'll expose the injury.

When I undo the victim's belt and zipper, I see that he is wearing long johns. In this heat? Strange. Did he simply still have them on after spending the night in the mountains? Did he expect that "the mountains" are always cold?

I pull back his pants and long johns, taking care not to move his spine, so I can see his right buttock. As I do, a huge, gaping cut on his buttocks opens and a teacup of blood spills out. I don't have time to study the cut, but it looks about three inches long and at least two inches deep, slicing deep into the muscle.

I immediately take my left hand and press it over the wound to slow the bleeding. Man, am I glad I put on gloves!

Drilled into every emergency medical service training is the commandment *always* to take body substance isolation, or BSI, precautions, such as gloves and possibly a mask, goggles, and gown. All body fluids must be treated as if they are infected with communicable diseases. During every practice scenario, EMTs are conditioned first to ask the instructor/evaluator, "Is the scene safe?" and then to hold up their hands and say "BSI" to the evaluators, to signal that they are wearing gloves (or that they would be wearing them if it were not a practice scenario). Forgetting to do so is an automatic failure. But in real life, especially in mountain terrain and cold weather, it's tempting—and sometimes appropriate—to postpone putting on medical gloves until they are needed.

Jim sees the open meat at the same time I do. His eyes widen and he quickly spins his head as if he is going to vomit. He slowly wags his turned head and says, "Man, I'm glad we didn't see *that* before you got here!" I am thinking precisely the opposite.

With my free right hand, I immediately key the radio on my chest, "Command Post, this is Nine hundred"—my current number, as commander. "I need a hoist." Back in the command post, the change in the tone of my voice, the urgency, comes through loud and clear.

Undermanned and Making Do

I am now in an unusual situation. Rescuers almost always work in teams of at least two—if one of us arrives at a trailhead and is ready to go in, we wait until another rescuer arrives. Teams of two ensure that if a rescuer gets into trouble, he or she has a partner. But this time the combination of altitude and weather dictated that I be inserted alone. I am now a solo, one-handed rescuer.

Fortunately, I have Jim and Carol.

I ask Jim to open my backpack and locate the trauma kit. Telling him how to retrieve my gear is tricky, because the buckles on my backpack are not obvious and I want specific gear from its many compartments.

I have Jim hand me several of the four-by-nine-inch "battle dressings." I then realize that with only one free hand, I can't even open them. Jim does.

Jim is great. Although he was initially shocked at the blood gushing from the wound, he is quite focused and follows directions perfectly. This is a new situation for me, too, since I've never had to work through an assistant.

Before I can apply the dressings, I need to get the victim's clothes out of the way. I explain to Jim where my medic shears are stored. Then, with one hand, I clumsily try to cut a vertical line down his pants. Seeing me struggle, Carol pulls the pants tight to help me cut them.

Jim then hands me a four-by-nine dressing. I quickly lift my hand away from the wound, lay on the dressing, and reapply my hand. We do this repeatedly. Each time I remove my hand, I see that the blood has completely saturated the previous dressings. We use all my four-by-nines, then all the smaller four-by-four dressings, with the same results.

Stopping bleeding is done using four progressive techniques. First, apply direct pressure to the site—I'm doing that with my hand. Second, elevate the injury, preferably above the patient's heart, but it isn't practical to raise his pelvis. The third technique is to use a pressure point to reduce the blood flow, but there isn't a pressure point that will stop blood flow to his backside. The final technique, a tourniquet, is rarely appropriate and won't work here. Since my hand hasn't stopped the bleeding, my next option is to improve the direct pressure by creating a pressure dressing.

I need more large dressings to control the bleeding. Unfortunately, all my clothes are "plastic," primarily polypropylene. They're perfect for the backcountry, because they breathe and don't absorb water. But I need dressings that will absorb blood. I look at Jim's somewhat grimy cotton T-shirt. I smile and ask if I can have it. Without a word, he whips it off, and I use my free hand to fold the shirt into an eight-by-eight-inch pad and put it over the wound, then slap my hand back onto the dressings to apply pressure.

Carol, who is also wearing a T-shirt, asks if I need it. These two are amazing.

I have Jim open up several rolls of Kling roller gauze. We pass the roll around the patient, with Jim reaching underneath the patient and passing it back to me, and me passing it over the patient to him. We do this repeatedly so that the bandage, rather than my hand, will keep pressure on the dressings. The additional pressure and the larger shirt help slow the bleeding, but they don't completely stop it.

Carol is helping during this process by stabilizing the patient's head and helping apply the bandaging. She isn't wearing gloves. I lean over to her and sternly whisper into her ear, "Do not touch any blood." The BSI warning has been burned into my mind; I don't want either of them getting sick by helping a patient. Many people would be freaked out by my warning, but Carol reaches into her fanny pack, pulls out two Ziploc bags, puts them over her hands like mitts, and gets back to work.

The patient repeatedly complains that he is thirsty. I check with Jim and Carol, who tell me they gave the patient almost two bottles of water since they arrived.

"Why won't you let me drink anything?" he asks yet again. My concerns about giving him water are twofold. First, the basic rule of thumb is to avoid given patients anything to eat or drink, because if they go into surgery, the anesthesiologist will want them to have an empty stomach. But if oral fluids will improve the patient's ultimate outcome, I'll certainly break this rule. Second—and this is of more concern to me now—if I give him something to drink in his current state, there is a chance he will vomit, be unable to protect his airway, and aspirate the vomit.

His excessive thirst sets off alarms in my head. Hypoperfusion is a medical condition that means the body is unable to provide its tissue with an adequate supply of oxygen. Laypeople call it "shock." The classic signs are altered mental status (my patient appears to "sleep" when I am not talking to him, and he knows only that "I fell"); pale, cool, clammy skin (he's got it); rapid pulse and breath (both are racing), and thirst. His pulse and respirations have increased markedly in the fifteen or so minutes I've been here.

I reach for his wrist to take his radial pulse—can't find one. Checking his carotid artery at his throat, I do feel a weak pulse. I take out my blood pressure cuff and stethoscope. The first time I get a systolic pressure of 70—that's very low. I try again and can't get a blood pressure at all! I then try to get a blood pressure by palpation, by feeling his wrist for a pulse rather than listening with my stethoscope. I can't.

He is getting progressively worse. His skin is gray, meaning his body is shunting blood away from the skin to the vital organs. He struggles to open his eyes when I speak to him.

The command post calls me on the radio. They are loading the helicopter and tell me the supplies they are bringing: the full-body vacuum splint, litter, and more ropes. Yes, we'll need all those things to lower him down the mountainside, but what I really need, desperately, is oxygen. Right now I am more concerned about stabilizing him than transporting him. At this point I'm unsure I can keep him alive.

One of the steps during the initial assessment stage is to formulate a general impression of the patient's condition. It doesn't take much pondering to conclude that I have a terrible general impression. This guy is neither stable nor dead.

I am losing him, and I know it. There is no way he can survive the six or more hours that it will take to carry him out—We need a hoist.

My patient has obviously lost a lot of blood. Normally, when someone bleeds onto a hard surface, a little blood can spread out and look like a lot. For example, if you spill a cup of coffee on a hard floor, it will spread out over a wide area. But this guy's blood has been trickling down through the rocks, so I can't really tell how much blood he has lost. And oddly enough, the human body can "lose" blood internally, too. A patient with a fractured femur can lose a liter or more of blood within his leg, and a patient with a fractured pelvis can lose two or more liters internally, even bleeding to death without spilling a drop.

The solution for blood loss is a blood transfusion. A stop-gap measure is an IV. The IV fluid can't carry oxygen the way blood can, but the increased volume may improve his sagging blood pressure and help circulate what blood remains.

Unfortunately, I'm in a terrible location to start an IV. We're on the side of a hill, on loose rocks. There isn't a place to set anything down. I have blood and grime up to my elbows. And truth be told, my IV skills are only fair at best.

As I begin setting up the tubing, a radio report announces that the hoist ship is available. That's great news. Now I am hoping he can survive until the hoist arrives.

Realizing that I am out of time to start an IV, I call the command post and tell them to have crews ready to start two large-bore IVs the moment the patient gets to the road. The experienced paramedics don't need me telling them what to do, but a heads-up on the urgency of getting the lines started may turn out to be the difference between living and dying.

During my assessment, I stopped when I encountered the life threat, the still actively bleeding laceration. Continuing the assessment, I have Carol support his head as we roll him slightly so I can assess his back. I ask him, "Does your neck or back hurt?" He says no. I run my fingers down his neck, pressing on each vertebra and repeatedly asking, "Does this hurt? Does this hurt?" He mumbles, "No, only my hip hurts."

Now that we have already cut off most of his pants and long johns, modesty is a lost cause. I reach my hand behind his scrotum and press on his "sit bones," continuing to prompt him, "Tell me if any of this hurts."

I move down both legs with my hands, feeling for anything out of the ordinary. In medical jargon, I am checking for DCAP-BTLS—pronounced "decap beetles"—the acronymic memory crutch reminding me to check for deformities, contusions, abrasions, punctures/penetrations, burns, tenderness, lacerations, and swelling.

I hear a lovely sound: the *whop-whop* of the inbound helicopter. Although it feels as though I've been here a long time, it has been only thirty or forty minutes—very *long* minutes.

I look up and see the brown DPS helicopter coming in. It lands on the green patch below us.

James Taylor and Tom Moyer climb out of the helicopter with additional gear. I'm glad to see that the pilot was able to bring in two rescuers—his first landing must have helped him determine that he would have enough power.

Tom and James hike up to me. James has the oxygen kit and begins to assemble the tank, regulator, mask, and tubing. Tom applies direct pressure on the wound to help slow the bleeding. I use my stethoscope to check lung sounds and again try to get vitals.

The radio reports that Life Flight has landed, offloaded unnecessary gear, rigged for a hoist, and is inbound.

I again hear the sound of an inbound helicopter. I can see a rescuer in a red jumpsuit hanging from a cable barely more than one-eighth inch in diameter, a few feet below the helicopter.

James is just completing assembly of the oxygen gear and is preparing to put it on the patient as the ship moves in to hover above us. James raises his eyebrows questioning whether he should continue. I wave him off over the sound of the rotors.

The nurse lowers the medic to the ground about twenty feet from us. I see his face—it's Paris. He grins. He obviously enjoys dangling from the thin cable. He is one of those easy-going people who never seem to get very excited. I suppose that for Paris, "very excited" just means his grin gets a little wider.

Once on the ground, he reaches into his backpack and pulls out a cervical collar. Tom, James, and I put it on and package the patient in the beanbag. Paris prepares the hammock-looking bag that will hold the beanbag. We load the patient into this bag and Velcro it shut. Paris gathers the loops on the ends of the dozen or so straps and holds them out away from his body. The helicopter hovers fifty feet above us, with the pilot and nurse working together to put the dangling hook into Paris's outstretched hand. Grasping the hook, he passes it through the loops and motions with his arm, and they rise into the air and fly toward the valley.

As this is taking place and I am packing up my backpack, James picks up my camera to take a picture of me. He tries to take a couple of photos, and yells over the sound of the rotors, "I don't think your camera is working." Unbeknownst to either of us, the switch on the camera has slid to the "movie" mode, and he took a short movie. It shows me packing my backpack, with the helicopter in the background and the patient and Paris dangling below. Although the movie is only a few seconds long, it is a fond memory of how we made a difference. Although our patients are usually either stable or dead, this one was neither.

To watch this very brief movie,
visit MountainResponder.com/Movies/JAJ

The DPS helicopter returns to pick us up and lands on the now-familiar green patch. The first "lift" takes the bystanders, Jim and Carol, to the valley below.

What Jim and Carol did for this complete stranger goes way beyond being a Good Samaritan. When they heard the rockfall they could have shrugged their shoulders and continued their hike. Instead, they scrambled across a cirque and found a stranger lying in the rocks. With only basic first aid training, they provided care before we arrived. And they were my voice-activated "hands" while I treated the patient. As rescuers, the rest of us were just doing our jobs, but these two were going way outside their comfort level to care for a stranger. Afterward they were personally recognized by the County at an awards ceremony. They truly deserved it.

Tom, James, and I carry our remaining gear down to the landing zone. The helicopter lands and we climb in. As I clamp on the familiar green headphones, Steve lifts the helicopter a few feet into the air and holds a momentary hover, then lowers the ship back down to the ground. Looking at me, he again shakes his head. I know the drill: we're too heavy. With a tired smile, I open the door, pull out my backpack, and kneel beside the ship. I give Steve the thumbs–up, and they fly off. *Damn*. Here I am alone, again. Of course, I'm thrilled that Steve figured out that he didn't have enough power during his power check and not after we were airborne.

Second-Guessing

It's odd—after a rescue, I never really feel as though I saved a life, because I've never done it alone. But I often feel that the *team* saved a life, and this is one of those times. Collectively, we did save the guy's life—and the most important part of "we" is Jim and Carol.

After being home for a half hour, I get a call from Deputy Burton. He tells me that the patient is at the hospital and has been downgraded to Delta, near death. I learn that he has multiple fractures, including a fractured neck and an open pelvis fracture.

The severity of the "cut" on his hip was caused by his fractured pelvis ripping through muscle and skin. It explains the signs and symptoms of blood loss—he probably lost a lot of blood both outward, through the cut, and internally. It also explains why it was so hard to control the bleeding—I wasn't controlling the bleeding from the laceration itself, but from a more serious internal injury. And finally, it explains why I was barely able to get a pulse—he was running out of blood.

It was frustrating to hear that I missed the neck fracture. I expect that the pain from his hip was so overwhelming that it distracted him from feeling pain in his neck when I checked it—if your leg is on fire, you forget you have a toothache. Still, I missed it. This is a good reminder that an assessment isn't an X-ray. It's also a reminder that in the field, I have to base my treatment on the mechanism of injury—a fall on rock—and not just on what the patient tells me. It won't be the last time I miss something, but I know I'll never forget this call.

A few days later, a deputy tells me that our "Lake Blanche patient" died. I am surprisingly depressed and again start second-guessing my treatment. *Crap!* Should I have delayed transportation a few minutes to get the IV started or to put him on oxygen? I find myself time-slicing the call, trying to find a way that I could have made a difference. The result of my second-guessing almost always results in the same conclusion: I did the best I could with what I knew. That doesn't keep me from questioning my actions, or lift me out of my funk.

After stewing for two days, I called Sergeant Griffith and asked him if he knew more particulars on our dead patient. He laughed and told me he had heard the same rumor, but he called the hospital—and our patient was alive and stable. It feels like a miracle—as if he were resurrected from the dead.

Steve assembles gear as Life Flight
flies the patient to safety
(Visit MountainResponder.com/Photos and enter Story Code LBR.)

Moss Ledge Struggle

```
This is a callout for S&R at Moss
Ledge in Big Cottonwood Cyn...meet
car 617...climbing gear needed...
respond on Code One Frequency...
case@47404..eom
 19:08 05/08/04
```

The call instructs us to respond to the Moss Ledge picnic area, about five miles from my house. There's a waterfall about a quarter mile from the picnic area.

Since the pager also asks for climbing gear, I assume the call is for a stranded climber above the falls. This has happened a couple of times since I joined the team. People climb up the rock walls on the sides of the waterfall only to discover that it is easier to climb up than down, or they get lost on the trail that goes along the mountainside above the waterfall. They then wander downhill until they find themselves cliffed out above the falls.

An Easy Callout

As I pull up to park behind a fire engine on the right side of the road, I see the short set of stairs leading up to the popular picnic area, where there are several picnic tables within a few hundred feet of the road. It should take less than fifteen minutes to get to the waterfall.

Two paramedics are sitting on the rear bumper of the fire truck. These guys are older paramedics, probably nearing retirement, whom I have worked with on numerous missions. They're agreeable guys who provide good care to our mutual patients.

One of the medics tells me that a sergeant with the sheriff's office has already headed up the trail. He tells me they were told that a patient fell and is conscious and alert in a pool of water.

I load up my medical gear—the infamous trauma kit—and the oxygen kit that includes an aluminum oxygen tank, tubing, mask, and the football-shaped bag valve mask, or BVM.

Gary Banks arrives at about the same time I do, and asks if I'd like him to get a rope. It's an unusual question—not whether we'll need a rope, but that Gary is asking it. He has commanded the team for the past ten years, and I've been following his capable, experienced lead. But recently he stepped down from the elected position, and I was elected vice commander. This is one of my first missions in this new role, with Gary, strangely, following my instructions. He has infinitely more experience—it somehow doesn't feel right.

I pull the bright-red beanbag from the back of my truck and lay it on the ground. Then, sliding out the orange two-piece litter mounted on a backpack frame, I put it next to the beanbag, at the side of the road. I ask the medics to have the next rescuer who comes in bring the vacuum splint, and the third person to bring the litter. Although we still need to locate the patient, I'm trying to make arrangements so the additional gear arrives in the order we will need it in. Gary will bring in a two-hundred-foot rescue rope, which may be needed to access the patient, and I'll bring medical gear, which may be needed to stabilize him. The beanbag and litter will be used for last LAST stage: transportation.

As I climb the concrete stairs toward the picnic area, my legs are complaining—I ran a 5K race this morning. In my more superstitious moments, I believe that the way to ensure a callout is to work out hard at the gym, compete in a race, or do some exhausting multipitch backcountry rock climb. It rarely fails.

There is no maintained trail, just a series of random paths, all of which head up the narrow canyon to the waterfall. The small stream is running high with tumbling water from spring runoff. The trail squeezes through the new season's foliage, goes over a brownish-red quartzite slab, and involves some minor third-class scrambling.

We meet Sergeant Powers of the sheriff's office on the "trail." He tells me it looks as though the victim fell while climbing a rock wall next to the waterfall. He is now in a pool of water near its base. Powers tells me to look for a girl in a white shirt, and another in a red shirt, who will direct us in. The route is straightforward, but having these spotters will help save time.

As we reach one of the first bends in the trail, I talk briefly to the woman in a white shirt; she points the way up the trail. After another bend, I see the woman in a red shirt, urgently waving her arms over her head. I'm a little surprised; this call didn't sound that serious. I pick up the pace. When we near her she yells, "Hurry, please!"—usually not a good sign.

I can feel my intensity and focus growing. I run through our equipment needs. Do we have everything we'll need? In my mind, I role-play the medical care I may need to provide.

The waterfall starts sixty feet above us. In the upper portion, the water drops forty feet vertically into a pool of water ten feet across. It then spills out of the pool and cascades down another twenty feet.

The steep, jagged rock walls on either side rise almost a hundred feet. I suspect that the victim fell from one of these walls, tumbled down, and landed in the pool.

Seeing a group of people standing in the pool, I scramble up the rock toward them. It's wet and slippery with dark-green algae. In arid Utah, even this small thin layer of green slime is noteworthy enough to warrant the name Moss Ledge. I move carefully up with my full backpack.

There are four young men, each grasping the edge of a blanket. The victim lies in their makeshift hammock. The men, teenagers really, remind me of old-school firefighters holding an open blanket below the window of a blazing apartment building.

I step into the calf-deep pool and feel the cold water fill my hiking boots.

The young men are holding the victim in this odd cocoon, keeping him just above water level. I assume they are his friends. It's clear that if they weren't doing this, the victim would sink into the frigid water.

I move directly to the patient and yell over the sound of the waterfall, "My name is Steve. Are you okay?" There is no response. His skin is ashen gray. He left eye is swollen shut. His right eye is open but staring blankly into the distance. There is blood on his face and trickling out of his left ear.

The Telephone Game

I press my gloved fingernail into his brow, hoping for a response to the pain. "Buddy, can you hear me?" No response. Nothing. I reach for my radio and talk over the sound of the waterfall echoing off the rock walls: "Command post, this is Nine-oh-one. I have a patient who is unresponsive to pain stimulus." It's an ominous sign that needs to be communicated immediately, before I continue my assessment.

If I weren't preoccupied with the condition of my patient, I might muse over how often, and to what a wide degree, the condition and situation that we find our victim in varies from the initial report. Only twenty minutes ago, I was told that this patient was "conscious and alert." And now I discover a patient who is unresponsive to pain. It reminds me of the "child in the river" who turned out to be climbers on a cliff. I will later learn that this call was relayed down the trail and passed to a motorist who had to drive to the valley for cell phone access. And I expect that the patient's conditions were probably passed through several people before getting to the Fire medics who passed the information to me. It reminds me a bit of the old telephone game, in which people whisper a word or phrase around a circle, until it arrives again at the originator, utterly unrecognizable. I have to marvel that the call got through at all.

As the young men continue to hold their friend above the surface of the water, I reach for his left wrist and feel for his pulse. I can't. Reaching up to his neck, I locate his Adam's apple and slide my fingers into the grove on his neck. I can't find a carotid pulse, either. I watch his chest for fifteen seconds; it rises six times.

He obviously has a pulse, since he's breathing, but a pulse too weak to be felt at the carotid artery is beyond ominous. I continue to probe his neck with my fingers. I feel a series of five very weak beats.

I call the vitals down to Fire: "I have an approximately twenty-year-old male with an intermittent pulse. Respirations are twenty-four. We need an air ambulance." I realize I don't have very much information at this time. He is obviously in critical shape, and transportation will be difficult.

Although the canyon walls look too narrow, I ask Gary if he thinks we can do a hoist from our current location. He says he's not sure, but he thinks we can. Although it's a narrow slot, the rock walls widen slightly down-canyon. I call the command post again and clarify that I want the hoist-equipped helicopter if it is available. Gary reminds me that we may run out of daylight.

The roar of the water makes communications difficult between Gary and me, let alone over the radio.

There isn't a dry place to put my backpack, so I leave it on my back as I resume my assessment. I start at his feet, which is unusual, but they are easier for me to access. I palpate his lower legs by clasping my hands around each calf and squeezing, moving them up several inches, and repeating. They feel okay.

I flip back a corner of the blanket that is covering his left leg, so I can palpate it. Immediately I see a huge deformity of his left thigh. I feel it and discover that the thigh muscle is bunched up, knotty and hard—obvious signs of a fractured femur. I can't access his pelvis, because he is too deeply sunk into the blanket. One of the men tells me that his left arm is broken, "here," pointing to the upper arm. I palpate both arms and confirm the apparently fractured humerus.

I palpate his abdomen, chest, and neck. As I palpate his bloody skull, I notice that the left side of his head, behind his ear, is deformed. I don't press it. A small trickle of blood is running out of his left ear—signs of a skull fracture.

Trying to obtain a SAMPLE history, I ask the blanket holders if their friend has any preexisting medical conditions or if he is taking any medications or drugs. They tell me they don't know him. This will remain one of those unanswered questions after the call. How strange that these young men stumbled across this fallen climber. Did they see him fall? Were they just scrambling up the rock when they found him lying in the pool? Was he fortunate enough to land with his face above water so he wouldn't drown? How many minutes passed between his fall and their arrival? Where did they get the blanket? Did they run back to a car? And what a coincidence that the victim was found by altruistic bystanders who were willing to help a stranger. I never got answers to these questions.

I am usually very punctilious about performing a good assessment. Memories of my missing a fractured pelvis on a backcountry sledder remind me that I'm as fallible as ever. But in this case I've seen enough. This young man has multiple, major traumatic injuries that I can't treat here. I can do a more thorough assessment when the time and location permit.

"Command post, this is Nine-oh-one. The patient appears to have a fractured skull, a fractured right humerus, and a fractured femur. I am stopping my assessment at this time."

My Mentor

This call is getting more complicated by the minute. I turn to Gary, who is still the only other rescuer on scene. Although we recently swapped roles, we both know he has more than a decade of experience on me.

He is setting up anchors so we can lower the patient. As he gathers the mechanical anchors, his small pack reminds me of Mary Poppins's carpetbag—small, but filled with everything you could possibly need.

Gary is the classiest person I've ever seen step back from a senior position and remain on the job. Most leaders have a rough time letting others assume the role they had, and I admire how well he has pulled it off. In the heat of this call, as I have been making decisions about medical care, the helicopter, and equipment needs, Gary has worked quietly beside me the whole time. When I ask him, "Am I missing anything?" he just says, "You're doing great, Steve."

Having him next to me in case I need help, yet feeling his steadfast support for me in the role of decision maker, is unlike anything I have seen in the world of business or rescue. He will always define for me what a mentor can and should be.

Treatment Decision

I need to get medical gear out of my backpack, so I shuck the pack and jam it into a crack in the rock wall. Pulling out the long purple bag, I begin to set up the oxygen.

Over the sound of the waterfall, one of the young men yells, "He stopped breathing! He stopped breathing!"

I hand the oxygen tank to Gary, who asks if I want him to set up the oxygen or to continue setting up the climbing anchors so we can get the patient down from our watery perch. We exchange glances, knowing that he can't do both at once. Our patient is struggling to live and needs oxygen, but we also need to get him off the waterfall so he can get medical care. This is the perennial challenge of caring for a critically injured patient—there just isn't enough time to do everything that must be done.

To Gary's question, I reply, "Finish the anchors."

I take a few steps across the pool to check the patient's breathing. It has almost stopped.

Pulling the oxygen kit out of the crack, I grab a bag of oral airways. "Oropharyngeal airways," as they are properly called, are hook-shaped tubes of curved plastic that can be inserted into a patient's mouth to keep the tongue out of the way and maintain an open airway. I insert the device into his mouth and slowly rotate it into position. As I do, I feel his broken jaw bones crunching.

He begins breathing again.

As I am working on the patient, one of the bystanders is desperately yelling, "Hang in there, buddy! Don't stop breathing! Don't stop breathing!" It's creepy. How often do you hear the sincere instruction that someone not stop breathing?

To the four guys holding the blanket, I say, "Tell me if he stops breathing or pukes." And I go back to setting up the oxygen bottle.

At about this time, Darren Hunsaker, the newly elected commander, arrives at the pool. He has the beanbag and begins unpacking and rolling it out. It looks like a red air mattress floating on the water. He moves it to the side of the pool, where the water spills over the edge.

My blue medical gloves are covered with blood. I pick up the oxygen kit with my bloody gloves and begin to set up the BVM. *Shit!* The mask is missing! I check the oxygen kit again—it isn't there. Then I see various items from the oxygen kit floating in the water.

The bystanders again start yelling that the patient has stopped breathing.

I check the pulse on his neck, trying to find any sign of life. There is none. I'm in a predicament. We can't do CPR in the water, yet he needs resuscitation immediately. I don't see an alternative. Since we can't provide care here, we must lower him. This puts us in the crappy position of providing transportation before stabilization.

Standing in the water, I pull on my climbing harness as Gary connects the rope to the beanbag.

Other rescuers have arrived, and as the bystanders maneuver the blanket to help us get the patient on the beanbag, they quickly strap him in with the wide straps and buckles. While he is being packaged, I listen with my stethoscope for a heartbeat and breathing. I don't hear either. I look helplessly at Hunsaker and ask if he wants me to attempt resuscitation here. He shakes his head and says, "We can't." I can hear the shared anguish. We both know we are losing him.

We don't take the time to make the beanbag rigid by vacuuming out the air, nor do we package it in a litter. We've run out of time.

Clipping the short leash on my climbing harness to the rope, I slide over the edge and drag the patient, in the sagging rig, behind me. Gary lowers us down the remaining section of waterfall. With water flowing over my legs, I struggle a little to keep from falling on the slippery footing, unaware that my cell phone is drowning in my cargo pocket. The floppy beanbag lets the patient's legs catch on something, and I have to pull on the end of the beanbag to keep his legs from folding under him. I also have to pull on the bag to make sure his head doesn't go in the water. The lowering takes only a couple of minutes, but it feels like eternity. As I am being lowered, the bystanders scramble over the edge. They meet me at the bottom and help carry him to a better location.

Calling It

I now have a patient who is not breathing, doesn't have a heartbeat, and hasn't had either for about ten minutes. He is unresponsive to pain, has multiple traumatic injuries, including apparent fractures to femur, humerus, jaw, and skull. His right pupil is fixed and dilated; the left eye is too swollen for me to open.

Disappointed and frustrated, I know that CPR isn't going to make a difference. CPR saves lives when the patient has had a heart attack, because it helps oxygenate the body until the patient's heart can be restarted with a defibrillator. But it won't restart a heart stopped by traumatic injuries, and we can't fix those underlying problems here. The knowledge is small comfort.

On the radio, I tell the Fire medics the situation: "The patient is in full arrest, has suffered multiple traumatic injuries, and has a major head injury." I pause, knowing the burden I am about to pass them. "Do you want to intubate him and attempt resuscitation?"

They remind us of our predicament by asking a question that lobs the ball back to me: "This is a traumatic arrest, right?"

The answer is obvious. Yes, this is a traumatic arrest. The implied bigger question, now in my court, is whether we want them to try to resuscitate the patient.

Laurie Jess is here with me. She's a fellow rescuer and a ski patroller, with strong prehospital medical skills. I discuss it with her and radio down the obvious though agonizing decision: "It's my opinion that we should not continue resuscitation." Their reply: "We concur."

We stand around and hang our heads for a few minutes. I can't believe we lost him. Neither stable nor dead. We rarely end up with a live patient we can't save; this is my first. His proximity to the road, the help of the bystanders, and his body's determination to fight to stay alive kept him going until we got here, but he was never close to stable.

We load the body and beanbag into the orange litter and cover him with a blanket. My patient is no longer a patient but a body.

This call was paged out at seven p.m., and it's getting dark as we don our headlamps. Although hiking in took less than fifteen minutes, transporting the body down the rugged terrain, in the dark, will take at least two hours. We connect ropes to trees to control the descent of the litter and its six attendants, who have now changed from rescuers to pallbearers.

Laurie and I clip into the litter and are lowered down the first serious pitch—a steep sixty-foot wall. Then, after several gentler pitches, the ground becomes level enough to carry the litter without a belay. The litter hangs from the ends of short "hero loops," which secure the litter to the attendants' wrists.

Most of my attention is focused on getting the body down, yet part of me has begun replaying my medical care and the chain of decisions I made. This is not a voluntary process that I can turn on or off—my psyche is doggedly trying to uncover what I could have done differently that would have saved this guy. Logically, I know that even if we could have started resuscitation in the pool, we couldn't possibly have continued resuscitation during this multipitch evacuation. But all the sound logic in the world doesn't ease the sense of frustration.

Fellow rescuer Chris Patch asks me how I am doing. I guess the disappointment shows, or maybe he's just attuned to the situation. I tell him I'm a little frustrated but okay.

As we near the road, I can see flames from fire pits in the picnic area. People are living life joyfully. Three or four couples are roasting marshmallows as we march solemnly by, carrying a corpse. I look up at Laurie, who shakes her head and says, "Isn't life weird?"

A few minutes later, we see television news cameras lined up at the parking lot. We stop for a minute to discuss how to get by them. We agree to put on our game faces and move quickly to the waiting Gold Cross ambulance. "Command Post, this is Nine-oh-one. We are one minute out."

Rescuers carrying the victim down the final pitch
(Visit MountainResponder.com/Photos and enter Story Code MLW.)

Reflecting

This callout will sneak into my consciousness repeatedly over the coming months. It will happen when I am slipping off to sleep, driving my car, or staring out a window.

I will replay the call, wondering what I could have done differently. Would he have lived if I had put him on oxygen before lowering him? Where was the BVM mask, and would it have made a difference? In my mind, I repeatedly arrange the perfect treatment progression, only to realize I would need to do thirty minutes of treatment in five minutes.

I know that this young man would have died if we were not there. It is undeniable. And in my heart I know that given his injuries, survival simply wasn't in the cards. But I *was* there, and it was my job to keep him alive.

He died despite the best care I could provide. I know that, and it soothes some of the sting. I knew that it would happen someday. My bigger fear is to have a patient die because I didn't provide the best care possible. It could be something I do, but it's more likely to be something I *don't* do—a failure of omission rather than commission.

Although I work hard to provide good care, I've made mistakes. So far, they have been minor and haven't affected the patient's health, but this death reminds me of the seriousness of what we do.

Heroes

Like Jim and Carol, the four young men encountered an injured stranger in the mountains who was neither stable nor dead. And like Jim and Carol, they did everything they could to help him. Unfortunately, it wasn't enough.

When I reached the command post, my sergeant and I agreed that these four men should receive a citizen's award from the mayor or sheriff. They never did. I suspect that the paperwork was lost in the bureaucracy.

The news of this guy's death made it into the newspaper; the selfless story of these bystanders didn't. It should have been on the front page. These young men, like Jim and Carol, were ordinary people who came upon a bad situation and, without being asked, rose to the occasion. I call them heroes.

Thirteen
Definitely Dead

Before joining Search and Rescue, I had seen only one dead person up close: a coworker who tumbled from the top of the Jackson Hole aerial tram when I was working there as a mechanic. One—that's it. By comparison, in the seven years since I joined the team, I responded to callouts with forty-three fatalities.

Someone Else's Tragedy

Though it's a little embarrassing to admit, the fatalities don't really bother me. Sure, sometimes when the situation makes me think about friends—as it often does—it stings a little. For example, when helping carry out a dead climber, I might think about climbing and my climbing buddies, and when I picked up a woman in Mill Creek Canyon who shot herself in the head, I thought about my daughters and prayed that their lives never get so bleak. But I don't shed tears or fret and mourn over the dead people.

When you read the newspaper the day after one of these accidents, it says something like "Climber Dies in Fall" or "Skier Killed in Avalanche." You might think, "What a shame!" or "That poor girl's family!" That you don't cry doesn't mean you don't have feelings. You just learned of a tragic death. And then you turn the page.

Being a witness to deaths that most people only read about in the paper or learn about on the evening news certainly changes my involvement, but it doesn't change the fact that they are dead. You may learn about an avalanche death on the evening news, whereas I helped locate the victim under the snow, dug him up, and carried the body out of the mountains. But the bottom line is, someone neither of us knew died in an avalanche.

In fact, rather than being depressed because somebody died, I'm often glad that I can be there to help the family by bringing out the body. And I must confess, while working with the dead I often experience a strange, almost euphoric exhilaration at the gift of being alive.

This doesn't mean I don't care—I do. I think about the victims' families, I'm sorry they died, and I wish they hadn't. I think about how death always seems such a waste. I hope the people I love don't die before their time. And then, like anybody else, I turn the page of the newspaper.

There have been times when I've had my fill of death, especially of people who go into the mountains to end their lives. I remember the man we found hanging by his neck from a tree, with a magpie perched on his shoulder. And there was the guy who tied weights around his torso, sat on the edge of a bridge, shot himself in the head, and landed in the river. That one looked like a murder to me, but what do I know? Once, when we were hiking in to recover a suicide, I was told, "He used a shotgun and there are brains all over." I had helped with several fatalities over the past month, and I decided that I'd seen enough. The newspaper headlines would suffice for me that day; I could skip over the details. I left the bagging to others and just helped carry the body.

Should I Hurt More?

Sure, there are times, seeing all this tragedy, when I wonder if I should hurt more. Hey, it's a little disconcerting to think that if it were me lying dead in the dirt, a bunch of strangers would just walk all around me, zip me into a body bag, and tote me off with no more emotion than they would expend on a double sack of potatoes.

Rescuers frequently use humor to cope with death. Gallows humor. We use it; so do cops and firefighters. Hearing us talk, you'd think we had no respect for the dead. But it isn't lack of respect—it's *separation*. The humor helps us treat the death at more of a distance, more as if we were merely reading a newspaper article about somebody else's tragedy. We are simply laborers doing our job.

On one of my early callouts, when I was concerned about how I would react to a dead person, my teammate Darren Westerfield said, "Steve, it isn't a person; it's a body." I've used the same line to coach other newbies through the experience. When I see dead people, they usually look peaceful. The life is gone; the soul has moved on to wherever souls go. We're just bringing home the empty container. Even when the body is pretty banged up, I usually feel that in some way, their spirit is at peace. Of course, the family is about to suffer like hell.

Friends and Family

But that separation, that arm's-length distance that we keep from death, dissolves when I see its effect on the people who loved the victim. I remember searching for several days for a missing photographer. It turned out he fell off a small cliff. When we finally found the body we were stoked. *Yes! Success! Mission completed! Whoopee!* As we rode back to the command post on ATVs, I'm sure we were wearing huge smiles. We had succeeded in our job after three days' hard work. As we neared the command post, I could see grim-faced people standing in the gravel parking lot in small groups. I mumbled, "Game-face, game-face," to my teammates as a reminder to wipe off the inappropriate grins and put on proper, solemn expressions. Just then the friends and family must have been told that he was found dead, because I could see heads drop forward. I saw a man put his forearm against a tree, his head on his arm, and start sobbing. I saw a man fall to his knees as a friend tried to catch him.

I recalled Laurie's remark as we brought out the dead climber from Moss Ledge, past the happy picnickers. This time the shoe was on the other foot. It's ironic how the same event, finding a body, can create a wonderful feeling of success for some people and the most horrendous, gut-punched feeling of despair for others. Surrounded by the palpable grief and shock of the friends and family, our mood shifted from the joy over hard-earned success to shared sorrow. A lump grew in my throat. Damn! Why did this young man have to die? Yes, Laurie, life *is* weird, isn't it?

There was the time we needed to tell a family that we had found their missing father and husband, dead. My lieutenant asked me to participate in the notification with him and someone from the Pastor Service. The family was from the Middle East, and the clergyman assured us that he had experience with people from different cultures. But none of us was prepared for the reaction. We walked solemnly in the front door, and in a somber voice, the clergyman told the family that we had found the man but that unfortunately, he was dead. The mother started a high, keening wail at the top of her lungs. She pounded her fists into her son's chest as he scooped her up in a bear hug and tried to carry her into a different room. A family friend dropped to the floor, moaning, and started to roll himself up under the rug. Keeping our game faces on, the lieutenant and I slowly moved to another room as he muttered, "*That* went over well." I remember thinking, as bizarre as their response might seem to a Westerner, if someone told me that someone dear to me had died, I'm sure that's how I'd be feeling inside.

Once I was sitting in the front seat of a deputy's SUV in a snow-covered parking lot, debriefing the friend of a dead avalanche victim. It was dark outside, and I had the SUV's heater blasting hot air in an attempt to warm the chilled twenty-four-year-old survivor. Patrollers from nearby resorts were still bringing the body down the mountain.

With a clipboard on my lap, I sketched a picture of the avalanche site as the friend provided details. The drawing showed the locations of the two men before and after the slide, the dimensions of the resulting avalanche, and clumps of trees that looked drawn by a child.

The skiers had been skinning laboriously uphill, with the soon-to-be victim in front, breaking trail. He turned around, looked at his friend, and said, "I don't feel good about this." The friend—who had tears streaming down his face as he repeated their conversation to me—asked, "Do you want me to go first?" "No," said the other. He turned his head and took a single step, and the slope fractured a hundred feet uphill and swept him to his death.

As the survivor recounted the story, he kept asking me what he should tell his friend's parents. His burden of guilt was palpable, and any answer I could think of to his question seemed trivial and blasé. I said, "Tell them how you feel." Venturing far outside my skills and training, I tried to comfort him as he began the long process of recovery. My throat was tight, and my eyes burned—not what you feel when reading a newspaper article.

Lost and Found

A body in the mountains gets treated differently from a body in the city. In the city, dead bodies are surrounded by yellow crime scene tape, and detectives come in with cameras and fingerprint kits. In the mountains, especially on big mountains like Denali or Everest, the dead must sometimes be left to the mountain.

```
This is a callout for S&R on a fallen
hiker...meet at the Park and Ride in
LittleCottonwood Cyn..respond on Tac
One..case#51267..eom
16:53 06/12/06
```

A teenage hiker called 911 and told the dispatcher he was in the mountains, bleeding, and didn't know where he was. He also said he didn't know where his friend was. All he remembered was that they were climbing a mountain called "Twin something." He gave the dispatcher a description of his car.

Deputies ran lights and sirens up both canyons while the dispatcher contacted the registered owner of the car, who turned out to be the victim's father. The father told the dispatcher that his son did take his car in the morning, but a different car. With the updated vehicle description, deputies quickly located it in Big Cottonwood Canyon.

```
Update on the S&R...respond to Park
and Ride in Big Cottonwood..respond
on Tac One..case#51267..eom
17:04 06/12/06
```

The caller's confusion and his statement that his head was bleeding were strong indications of a head injury. We would later learn that in addition to his fractured skull, both his wrists were fractured and both knees seriously injured.

Realizing the urgency, Darren Hunsaker contacted the victim's cell phone carrier and discovered that he was connecting through a cell tower 21.6 miles away. Using mapping software, Darren drew a circle with a 21.6-mile radius—it passed just below Twin Peaks.

Life Flight was called and, using Darren's coordinates, quickly spotted the victim on a snow-covered mountainside. His friend, obviously dead, was spotted several hundred feet downslope. With Paris on the hoist, Life Flight was able to lift the survivor to safety before the sun set.

Retrieving his friend's body wouldn't be so easy. Rescuers who saw the dead teenager said he landed on top of a natural snow bridge at the base of a cliff band. They could see water flowing under the bridge. The slick rock on the cliffs above the victim was covered with snow. Rescuers were concerned that the area was susceptible to a glide avalanche, in which the snow can slide as a unit on the rock slab, much as a steak will slide on a tipped plate.

To recover the body, one rescuer would need to go on belay, climb onto the small snow bridge, and attach a rope to the body. The belayer could protect the climber if the snow bridge collapsed, but both rescuers would be exposed to the avalanche hazard from above.

It was obvious that exposing the rescuers to the avalanche hazard, especially after the snow had been warmed all day by the summer sun, was risky. Although the chance of the snow sliding spontaneously was small, the consequences would be catastrophic.

I explained to Sgt. Todd Griffiths, supervisor of the Search and Rescue Team at the time, our concerns about the avalanche hazard. I told him we wanted to wait until the morning. If it could be done safely, we would then have Life Flight pick the body off the snow while the rescuer remained on the cable, and if that wasn't successful, we would send in two rescuers and remove the body quickly. Yes, there were still risks to the rescuers, but a quick mission during daylight, after the snow had all night to cool and firm up, was a risk we were willing to take.

I was standing next to Todd as he used his cell phone to explain our plan to his supervisor—I believe it was the watch commander. The supervisor couldn't believe we were going to leave a body unattended on the mountain and go home for the night. I can understand that perspective if the body were in a field somewhere, but I was looking at it from a mountaineer's perspective—the kid was already dead, and the rescuers weren't. But here was Todd's superior, insisting that he send in rescuers to spend the night with the body.

"Sir, we can't risk the lives of rescuers to—" Todd said before the voice on the other end interrupted him.

Covering the mouthpiece of his phone, he looked at me and shook his head in exasperation.

As commander, my primary responsibility was to look after the safety of the team. When it became obvious that Todd's supervisor was insisting that we have rescuers risk their lives to spend the night with a dead body, I pulled my one remaining card. With Todd covering his mouthpiece, I said, "Todd, don't tell him unless you have to, but if I am ordered to have rescuers spend the night up there, I will resign right here in the parking lot." That was the one and only time I ever threatened to resign, and I was absolutely serious. If a leader won't resign over the safety of his team, he isn't fit to lead.

"What if an animal drags the body away?" he asked in a low voice.

I grinned at him and said, "No way!"

Todd has a laid-back personality, and I had never seen him get angry, but he really went to bat for me on this one. His jaw tightened, he raised his voice, and without mentioning my ultimatum, he told the person on the other end of the phone that he would not send in rescuers tonight. I don't know if he had ever been in a position where he had to refuse an order from his superior—that really isn't how the sheriff's office functions—but he was visibly upset. He may have lost points with his supervisor, but he gained the team's respect.

After hanging up the phone, he looked at me and said, "Steve, if that body is gone, I'm history."

"It'll be there," I assured him.

The page came out at seven the next morning.

```
Search & rescue callout on a recover
of a fallen hiker at Twin Peaks. Meet
at the park and ride mouth of big
cottonwood cyn. Cs#06-51267. eom
7:11 06/13/06
```

I climbed into the helicopter to get a better view of the terrain and the avalanche hazard. I wanted to understand the balance between the risks to rescuers on the ground and the risks of a hoist operation.

As we flew up the canyon, I could see a layer of snow hanging precariously on the smooth rock. We then flew along the cliff band below the slope to gauge the difficulty of accessing the body on the snow bridge—the body wasn't there. Circling around, we came in low and slow, with four of us scanning the snow. Nothing. We continued this process, flying up and down the canyon looking for the missing body, until the pilot told me we were low on fuel.

Keying my mike and knowing how my words would be received by Sergeant Griffiths, I gave him the bad news: "Command, this is Nine hundred. We are unable to locate the body."

There was a long delay. He had stuck up for the rescuers, and I was letting him down.

Breaking the silence, I told him I wanted to bring in the Life Flight crew who spotted the body yesterday, to see if they could locate it.

He agreed.

About an hour later, the Life Flight crew from yesterday, with a fresh load of fuel, flew the canyon. They discovered that the snow bridge had indeed collapsed and partially hidden the body. A rescuer retrieved him using the hoist.

Life Flight arrives with the bloodied survivor
(Visit MountainResponder.com/Photos and enter Story Code TPF.)

"I Think His Head Just Fell Off!"

I will never forget those words. It all started when my pager
sounded as the sun was setting.

```
THIS IS A CALLOUT FOR SAR..1595 W
NORTHTEMPLE..MEET CITY PD..ON A BODY
IN THE JORDAN RIVER..CASE#12111..
RESPOND ON SO REQUEST ONE FREQUENCY..
EOM
17:52 01/31/03
```

The pager's words "body in the Jordan River" mean it is
an obvious fatality—he's dead; no need to hurry.

Driving to the location, I notice how different this area is
from our typical rescue terrain. Rather than my usual natural

surroundings, I am hemmed in by aging steel buildings in an industrial zone. A large electrical substation buzzes nearby. The area is fenced off with chain-link fences topped with barbed wire. This isn't going to be a mountain rescue.

Water spills over a low-head dam that crosses the river. A narrow walkway above the dam is supported by a dozen or more vertical steel pillars. The water has caught our victim, pressing him into one of these pillars.

Gary Banks is the commander when we receive this call. He and a handful of police officers are on the walkway, inspecting the situation. The submerged body can barely be seen through the wooden pallets, branches, and other debris that have piled on and around him.

Gary asks me to pilot the twelve-foot Zodiac boat. He selects two other rescuers, Alan Erdahl and Rob Hunter, to join me. We are to secure the body first and then figure out how to extract it.

Gary has the remaining rescuers set up a highline across the river. This consists of a rope strung between chain-link fences on either side of the river. Normally we would string the highline from trees—another reminder that we aren't in the mountains. Two more ropes are attached to this rope with pulleys. The assembled rig allows two teams, one on either side of the river, to work in concert to shift the boat left or right and to move it up- or downstream.

The moon isn't out, but we do have the light from the power substation reflecting off the river's surface. A few hundred feet upstream, a half-dozen rescuers carry the inflatable boat and outboard motor to the river's edge. The three of us put on our personal flotation devices and climb in.

I'm in the rear of the boat, steering the outboard motor. It's strange driving a boat in a moving river, even one this slow. When we're going downstream, I don't need to apply any throttle and the boat cruises along at a decent clip, whereas going upstream requires quite a bit of throttle to make any headway. And turning around requires me to constantly adjust both throttle and steering based on the current.

The three of us kneel on the floor of the boat, crouched with our heads down near the inflatable tubes as we pass under two low bridges. They slide silently above us.

I drive circles in the hundred-foot-wide river, trying to hold our position as rescuers on either side finish setting up the highline. We then attach a rope from the highline to the stern of the boat.

I can now give instructions over the radio, such as "River left!" or "River right!" and have our now-tethered boat move appropriately. Using these commands, we position the bow of the boat a few feet upstream from the victim.

Rob points to the victim's hand, which is all there is above the surface. It's black and almost fleshless from being exposed to the air. The rest of the body, under the surface of the water, has been relatively well preserved by the cold water and lack of oxygen.

Leaning over the bow, Rob ties webbing around each of the victim's legs. I sit on the inflatable boat's bouncy pontoon near the stern.

Just as the pillar trapped the body, it has also been trapping river debris. Now, with the victim's legs secured to the boat, Rob takes a twenty-foot metal pole with a grappling hook on the end and removes the collection of lumber, forklift pallets, branches, and plastic bags that are covering the body.

The legs, which surfaced after they were connected to the bow of the boat, are a bloodless catfish-belly white, with grotesque gashes that look like cuts in the legs of a wax mannequin. The head, once Rob removes the debris covering it, is equally white, with a ghoulish-looking face that reminds me of a Halloween mask. It is lying horizontally across his shoulders. We can't tell whether he has a broken neck or the head is detached from the body—either way, it doesn't look human.

After almost forty-five minutes of removing debris and waiting for a wire litter, which we will submerge to retrieve the decaying body, it shifts.

"Shit!" Rob says. "I think his head just fell off!"

He calls up to the people on the walkway above us, "Can anybody see if his head is still there?"

I shake my head and tell Rob, "Most people go their entire lives without hearing a statement like that."

Rob looks at me, smiles wryly, and says, "You've just been hanging around the wrong people."

A crowd of cops and firefighters has been gathering on the walkway eight feet above us. Their proximity to the recovery gives them the opportunity to second-guess our every move. It seems as though everyone has a suggestion on how we should remove the body, and everyone thinks we're taking too much time. Such is life in the spotlight.

The observers shine their lights on the victim and view him from several angles. After a few minutes, they call down to us that they think the head is still there.

We discuss tying ropes around the head, but decide that it will be too difficult with the head under water and the body pushed into the pillar, and that we are just as likely to lose the head while trying to secure it as while simply moving the body and hoping for the best.

Rob ties several more ropes to the body, and he and Alan connect the ropes to the bow of the boat. The limp, gangly body, connected to all these lines, reminds me of a bizarre life-size marionette, with us its puppeteers.

I'm wearing the basic swiftwater rescue gear: a personal floatation device, helmet, headlamp, knife, and throw bag. I'm also wearing blue, Nitrile medical gloves to protect me from the "biohazard." During tonight's recovery, I realize that I can't practice BSI—containing and controlling the spread of whatever microbes are growing on and in the dead flesh. If I assume that *any* contact passes germs from the victim—and it can—it is simply impossible to control the spread. When

Rob touches the body while tying webbing around it, he gets the nasties on his gloves. He then uses the grappling hook to remove debris from the body and passes the grappling hook to me, which presumably deposits them on my gloves. I then pull on the rope that has been around the body, which puts more of the bugs on my gloves. When I then touch the motor's throttle, use my radio, or hand Rob a carabiner, I continue spreading the invisible critters. Although I don't actually *see* biotic goo slimed onto any of this, I am psychologically aware of the inescapable spread of the biohazard. *Yuck.*

The three of us heave-ho on our marionette's ropes in an attempt to pull him off the pillar and out of the remaining debris. Using my radio, I give the command "Up river!" and the on-shore rescuers pull on their ropes. As they do, the body rises slightly to float just under the surface.

As the boat is moved farther upstream, the body surfaces and releases the gag-inducing odor of decomposing flesh. I swallow awkwardly and turn my head.

Starting the motor, I disconnect from the ropes, and ferry the boat to a patch of dirt near the side of the river, where several other rescuers meet us. Gary confirms that the head hasn't fallen off. *Whew!* I'm relieved that we haven't lost that potential trove of evidence.

Rescuers hold the boat onshore while others wade into the shallow water to retrieve the body.

I step out of the boat and walk away. I want nothing else to do with this rotten body. I walk from the riverbank to a steel walkway that is a foot wide and four feet off the water. Walking on it requires my full concentration, and like a gymnast on a balance beam, I focus my eye on its far end. The concentration distracts me from the work going on behind me. Good.

When I reach the other side, I turn around. Several people are rolling the body into a wire litter that they have submerged in the shallow water. As they transfer the body from the litter to the body bag, it splits open and releases its contents. Rob, who is helping with this task, spins around and heaves—I don't know how he lasted as long as he did. The smell is unimaginably vile.

When I get to my truck, I grab several alcohol wipes from my med kit and use the little squares to clean my hands and radio.

When the body has been bagged and moved to the medical examiner, I give Rob, Gary, and Alan a ride back to their vehicles. On the way, we discuss the stench of the body and the puzzling fact that we still smell it. Then Rob points to something on Gary's thigh, starts laughing, and says, "Look, you've got it on your pants!"

When I get home, I take off my clothes in the garage. Naked, I carry them into the house and put them in the washing machine. I also put my river rope, fleece jacket, and anything that can be washed into the machine. If this stuff spreads on contact, *everything* is contaminated.

Then I go upstairs and take a shower. Although I'm pretty sure I don't have any bio on my body, I scrub myself twice from head to toe. I even push soap under my nails and wash my hair twice. As I'm doing this, I know that the scrubbing is psychological, but I still feel as though I can't get clean.

As I slide into bed, I still smell the stench. I know it isn't on me or in the house, but the odor has burned itself into my nostrils and my mind. I'll have "olfactory flashbacks" for weeks.

It turns out the victim has been missing for five weeks. Silently decomposing while waiting for us to retrieve it.

A Beautiful Departure

Most of my days are not much different from the day before, or the day before that. Today will be different.

```
S/R CALLOUT FOR ASSIST WITH DOWNED
AIRCRAFT, CARS TO COME ON DUTY ON REQ
ONE FREQ AND MEET CAR 900 AT SPECIAL
OPS AT 1000. EOM.
```

I had read in the paper that a small plane crashed in the mountains in a nearby county. It was located yesterday when the weather cleared. This page is requesting that we assist the other county by providing "mutual aid."

On a hillside is a cooked plane. Burned to a cinder. It slammed into a rock outcropping in bad weather. From a half-mile away I can see the small black smear on the hillside.

We ride ATVs to the crash site. There is a burned fuselage and a tail, but only one small section of wing. I assume that the rest of the plane disintegrated on impact or in the subsequent fire.

The leafless scrub oak has been sheared off where the plane acted as a hedge clipper moments before meeting the rock.

The plane must have been moving fast; one of the three-bladed propeller assemblies is fifty feet up the hill. A man's torso is thirty feet away. I can tell it is human, because it has a head with a face. It isn't easy to recognize the torso. It's maybe three feet wide—far too wide for a human. I'm not sure how it got so wide—the back might have been filleted. I can't tell if there's a pelvis. There is an arm, with a waxy-white hand. We don't see the other arm. The heavy gold chain necklace looks strangely clean, stark against the red meat. Most of the clothes, and skin, have been stripped.

One of the old-timers on the team is nonchalantly picking up bits of flesh. He stacks the scraps in his gloved hand until it is full, then dumps it into a body bag. There are three other rescuers picking up the pieces; one tosses a kidney into the bag while another gathers skull fragments.

I am standing with the rest of the pack, about fifteen others, some forty feet from the crash site. We are all wearing white Tyvek suits to protect us from the biohazard. Some of us, including me, are also wearing Tyvek booties.

Watching the body collectors, I wonder how I would react to picking up pieces of human. I see others doing it, but I'm afraid I might get sick if I try. It takes me only a few minutes to realize that I want to try this, to experience it and to see how I will react. If I puke, at least I'll know this task isn't for me.

As I walk away from the group of onlookers, one of my newbie teammates asks, "Are you *really* going to do this?" I reply, "I won't know how I'll react unless I try it." I wonder if I sound braver than I feel.

I am soon in the thick of it, but avoiding picking up the human scraps. When we roll the torso into the bag, I help by holding the bag open. Others grab bones or meat on the torso as they roll it.

I've done a good job of avoiding actual contact with the remains, when a rescuer from the nearby county calls to me that there is some brain by him. He is in a nonparticipant group about twenty feet from the recently packaged torso. Since I'm wearing a Tyvek suit and gloves and he isn't, I walk over to inspect his find.

He jokes that he can kick some dirt on it and we can pretend it isn't there, but it's a quarter-pound chunk of brain. A piece of clear plastic windshield is stuck in it. Not wanting to touch it, I try lifting the slab of brain with a stick. I'd have as much luck lifting Jell-O—it bisects in two. I reach down and pick it up with my gloved hand and walk it to the body bag. I struggle to unzip the body bag with my one free hand while holding the slippery brain in my other. After re-zipping the bag, and seeing that my hand is now contaminated, I realize that I have lost my virginity. To think that as of a year ago I had never even touched a dead body, and now I'm tucking brains in a bag.

Darren Hunsaker asks me to go on the other side of the plane and help Jon Blackburn with a leg stuck under the wreckage. I can see a protruding booted foot under my side of the wreckage. It reminds me of the Wicked Witch of the East's feet protruding from under Dorothy's house. Jon, on the other side of the fuselage, says he can see the head of the femur. Several of us raise the plane as Jon drags the leg out by the boot.

There isn't a body bag nearby, so I tell Jon I'll carry the boot if he'll carry the femur. He laughs. As I pick up the boot, the rest of the leg barely moves. It is held together with skin and muscle and sinew, but the bones are crushed. We carry the leg to a body bag, me holding the boot and calf and Jon holding the end of the femur. As we slide the parcel of human tissue and boot into the bag, I belch and hope I don't vomit. We find what appears to be a foot not far away. It, too, goes into a bag. We never find the other leg or arm.

I open another body bag for the woman. Actually, I don't know how we can call this a woman—I can't be certain that it's human. It is a charred stump, about a foot long, with bones poking out at various angles. Jon inspects one of these, a one-inch-diameter bone that, with its marrow center burned out, it looks more like a yellow-gray pipe. He thinks it's a humerus, an upper arm bone. Others think it's a femur. With nothing connected to anything else, it's that hard to tell. We load the piece of human into the bag.

Despite the horrific appearance of the bodies, they died a painless death. Really, it doesn't look like a bad way to go. Husband and wife, flying the plane they love, and an instant death. In all the violence and gore of the remains, it looks like a peaceful, beautiful departure with a loved one. A few decades premature, but still beautiful.

It takes a few hours to finish our macabre scavenger hunt. As I remove my Tyvek suit, I pay strict attention to the clean versus dirty side. I don't want my contaminated gloves touching the inside of my suit, or the outside of my suit touching me. After a short ATV ride back to the command post, the first thing I do is wash my hands. Twice.

Hamburgers are on the grill. The meat looks not so very different from the pieces we just picked up. We are, after all, meat. Several of us eye the food nervously, and a few people mumble, "Not today" and skip the free meal. I consider skipping lunch, but it's five o'clock and I've had only a few energy bars all day. The burger tastes great.

Photograph by Christopher Patch

Rescuers confront the mangled plane
(Visit MountainResponder.com/Photos and enter Story Code UPC.)

Fourteen
Mass Casualty

My pager goes off at 7:15 on a frigid Saturday evening in January.

```
S&R Callout...Injured hikers on Mount
Olympus. Stage at Mt Olympus
Trailhead 5600 S Wasatch Blvd. IC is
614 on case Co06-6224. Wilderness
Rescue Gear required. Eom
19:15 01/21/06
```

I put on my winter clothes: long johns, fleece, and Gore-Tex. It takes me only a few minutes to drive to the trailhead.

I get a quick briefing from the deputies staged at the trailhead. A group of seven members of the Korean Alpine Club of Utah—a club we haven't heard of—were climbing Mount Olympus. Three or four have fallen and are injured. A member of the group called a friend in Salt Lake City, who called 911. It sounds as though the climbers on the mountain may also have called 911.

I speak to the friend on my cell phone. It's difficult to understand his heavily accented English. He tells me someone fell four hundred feet but "they are alive." I am a little confused by the conflict between his statement that "someone fell" and the plural "they are alive." I ask how he knows "they" are alive and he tells me, "Friend sees them move." He goes on to tell me that the seven climbers range from forty to sixty years old.

I call the sheriff's office dispatcher, who tells me a slightly different story. She was told that three climbers fell five hundred feet and that their ages are between thirty and forty. She says the caller told her that he "saw them crawling."

There is also confusion about their location. I am told both the front and back of Mount Olympus. It's a big mountain, and the difference could mean many hours just in locating them.

This is the typical confusion early in a call. I accept that all we really know is that one to four people may have fallen up to five hundred feet, somewhere on Mount O.

If I consider the worst-case scenario, that four people really have fallen five hundred feet, this is going to be an enormously challenging mission. It will be challenging because any more than one patient in a backcountry rescue is always a challenge. Also complicating the rescue is that it's January and the mountain is blanketed in snow. Will it really be possible for us to locate, access, stabilize, and transport four patients back to the valley before they die? This assumes, of course, that they're still alive.

An MCI, or "mass casualty incident," is an event with "multiple casualties which requires resources beyond the scope of the responding agency." That's FEMA jargon for something big and bad. This call sounds as though it fits that definition, because it is unlikely that we can complete this mission without assistance.

The friend who received the phone call from the climbers on the mountain arrives at the staging area. I talk to him and try to clarify the story. Nodding his head vigorously, he is adamant that three or more people fell four hundred feet. He desperately wants to help his friends on the mountain. I try to explain that we need him to stay at the command post, but he's anxious to help and wants to go on the mountain. Even if I can convince him that he will be of more help at the trailhead, I can't spend any more time dealing with him. I ask a deputy to make sure our "witness" doesn't go onto the mountain. This will become a challenge for the deputies, who eventually have to threaten to restrain him.

Helicopter Insertion

If we attempt to search the mountain on foot, at night, in winter, the sun will almost certainly rise before we locate the climbers. We call for a helicopter.

The Life Flight helicopter with a crew of three arrives a half hour later. They circle the mountain and report that they can see the party's headlamps on the back side of Mount Olympus, seven hundred feet below the 9,026-foot summit. Our victims have been located.

Life Flight's protocols prevent them from using a hoist in the dark, for obvious reasons. On foot, it would take at least seven hours to reach the climbers. We're hopeful that the helicopter can help us reduce that time.

Deputies close Wasatch Boulevard so the helicopter can land on the road. Along the road are power lines, now hidden in darkness, which the pilot must avoid. After landing, he shuts down his engine. The nurse, paramedic, and several rescuers help unload medical equipment from the helicopter to lighten the load and make room for rescuers.

James Taylor and I are in the parking lot, preparing to be flown in. I scramble down the steep dirt bank from the parking lot to the helicopter several times—first to consult with the pilot, Denny Patterson, and several times to my truck for additional equipment. It's difficult packing for this call, because we will need ropes and climbing gear to access the patients, medical gear to stabilize them, and winter camping equipment to spend the night on the mountain—and all the gear must fit in our backpacks.

Before getting into the helicopter, I hand off operational command to Darren Hunsaker. As the former commander, he's the right guy to handle the complexities of this mission.

I climb into the back seat of the helicopter, facing forward. James sits across from me in a rear-facing seat.

I strap on a helmet with night vision goggles. They look like small binoculars that are mounted on a hinged frame on the brow of the helmet. A small shot-bag weight on the back of the helmet balances it so my head won't be tilted forward by the goggles' modest weight. They are currently in the upraised position above my forehead. The Life Flight paramedic, Kyle Lavender, reminds me which button to press to lower the binoculars into position in front of my eyes.

As the ship lifts off the ground into the darkness, James smiles at me and gives me the thumbs-up sign. I smile back. Another adventure begins.

I reach up, press the button to lower the goggles into place, and the entire apparatus drops into my hands. Crap! I hope the rescue doesn't go like this! I move the goggles back into position and reattach them to my helmet.

The night vision goggles illuminate the mountain like daylight, although it is a bichromatic version of daylight, made up only of black and green.

After circling to gain several thousand feet of elevation, the pilot makes a flyby on the backside of the mountain. I can see bright green dots—the goggles' perception of the climbers' headlamps—in a steep couloir below us.

I lean forward to James and shout over the noise of the helicopter, "It looks like significant avy hazard above them." I'm referring to the steep snowfield above the green dots.

James nods his head and says, "Did you bring a beacon?" *Crap.* I shake my head. In the confusion of packing, we both forgot to put on our avalanche transceivers. This is especially embarrassing, because I publish a website about avalanche rescue and transceivers.

The pilot then flies back to the ridge that lies south of the south peak. Normally, helicopters can't land here, because a windstorm several years ago knocked down a bunch of trees, making landing impossible. But this winter the treefall is buried under snow, and the pilot is willing to consider a landing.

We approach the ridge from the east. As we do, the fresh powder billows up from the ground, and the pilot backs off. He positions the ship about two hundred feet from the landing zone and slowly descends toward the ridge. When we are about fifty feet from the landing zone, he applies strong power. The helicopter shudders, and I hear the loud *whop-whop* of the blades cutting through the cold air. The swirling snow almost envelops us in a whiteout, but he pulls back just in time. He repeats this process several times, each time approaching from a slightly different angle.

I'm not sure what's going on. I see us come in for a landing, I feel the pilot open the throttle, I see the clouds of snow almost blind us, and he pulls away. It feels as though we are repeatedly aborting the landing. After a few attempts, I realize that he is using the wind from the rotors to blow away the loose snow, so we don't land in a whiteout.

He then positions the helicopter about fifty feet out from the intended landing zone and hovers—the bird is hanging so absolutely motionless, it feels as if we are parked in midair. He takes out a laser pointer and points the red dot out the front windshield, at various features of the landing zone. He has a conversation with the paramedic in the left front seat. I can't hear them, but they're talking back and forth. The paramedic takes out his laser pen and draws a red circle on the snow.

The pilot gradually moves us toward the LZ, rotating the ship a few degrees. The paramedic opens his door and looks behind the ship. We're ten feet off the ground and almost motionless. The pilot turns the ship a few more degrees and slowly lowers it until the small skis, which are attached to the three tires, just touch the snow. As he does this, I press my hands together in the prayer position and look at James. We exchange nervous smiles. I know there is hazard in this landing. I also know that I have almost no control over the situation, other than to tell the pilot I don't want him to land. But that isn't the case. I *do* want him to land. James, sitting backward in the plane, seems happy to be on the adventure, and oblivious of the needle-threading going on behind him. He later tells me he wasn't concerned until he saw the look on my face.

The pilot slowly lifts the ship into the air and backs away once again to park in the dark sky.

Again pilot Denny and paramedic Kyle take out their laser pointers. I can see the red dots bouncing around on the snow. They seem to be focusing on a little bush protruding from the snow, which will be near the tail rotor when we land.

The pilot again comes in for a landing. This time he turns the ship a little more and again lets the skis gently touch the snow. I can feel the crunching of the skis as they sink into rotor-scoured snow. The pilot turns his head and nods at us.

The paramedic gets out the left-front door, walks around the front of the ship, and slides open our door. We climb out, pulling our big, heavy backpacks behind us. They sink into the snow. We close and latch the door behind us and crouch down next to the right door while Kyle returns to his seat. Denny looks out the window at us and makes eye contact before lifting his ship into the air.

The blasting wind from the rotors whips and swirls the snow around us, so that it finds every tiny opening in my winter clothing. Although I'm wearing ski goggles and a parka, it still sneaks in through the hood of my jacket and into the gaps between my gloves and sleeves.

As our taxi heels away from us and toward the city lights, I'm thankful for Denny's piloting skills. He was the right man for the job. The landing was cautious and precise—it also got my heart pumping. I feel safer with my feet on the ground, even when the ground is a snow-covered mountain.

It is a strange, dreamlike feeling to be in a city of a million people and, a half hour later, to be in the silence of the mountains, thousands of feet above it all. I love it. It's a little after nine p.m., and our adventure is just beginning.

Climb to the Summit

The light from the city below illuminates the sky, and the half moon reflects off the snow. James and I hoist our packs onto our backs and begin the six-hundred-vertical-foot scramble to the summit.

As we climb toward the summit, our pagers go off again. More rescuers are being requested. I know these folks will be helping with logistics. They may also be brought in to back us up in case we need help on the mountain.

```
S&R needed on missing hikers Mt
Olympus meet at the trailhead to Mt
Oly com 10-8 on tac 1 IC is car 600
cs 6223 eom
21:58 01/21/06
```

Over the radio, we hear that a second team, Dan Smith and Keith Sauter, is being flown in. The pilot, who now has a better understanding of the terrain, will select a new landing zone a short distance from ours.

The climb to the summit takes James and me about two hours. This is normally an easy scramble up the craggy rock, but the steep snow-covered rocks and our bulging backpacks more than double the time and exertion.

The top of the summit is approximately twenty by thirty feet. From it, we can see the entire 750 square miles of Salt Lake County and all the way up to the Great Salt Lake. It's a spectacular view.

The Korean Alpine Club team started their climb where the command post is now parked. They hiked the first few hours and then snowshoed for another hour or two to reach the ridge where James and I landed. Then they climbed the same steep snow to the summit and enjoyed the same view before things went wrong.

I call down to the command post to tell them the snow on the summit is level and firm and that we think a helicopter may be able to land here. Normally, the rocky summit is too uneven and slick to use as a landing zone, but if a helicopter can land here, it will save other rescuers the two-hour climb that James and I just made.

I'm glad James Taylor is my partner on this mission. He is the real deal. The son and grandson of mountain explorers, he grew up in the mountains east of Salt Lake County. Although he has a real job as the headmaster of a small private school, his passion is guiding less experienced climbers throughout the world.

From our perch on the summit, the mountain drops three hundred feet to a small saddle between Mount Olympus's two peaks. The narrow saddle then climbs three hundred feet to the top of the north peak, with the terrain dropping steeply off from both the saddle's flanks. To the west, the slope descends to Tolcat Couloir, the city, and our command post. To the east, the steep couloir descends to our waiting victims.

Given their current location on the back side of the mountain, it looks as though they downclimbed the steep north face to the saddle between the two summits. That is the route James and I will take.

We agree that we should set fixed ropes down the face so that we, as well as the inbound rescuers, can rappel to the saddle between the peaks. I tell James that I want the rappel stations to be absolutely bomber, with obvious tie-in locations at each station that the rescuers can use as they move from one rope to the next.

James is all for it. He's a cooperative guy, and I know he's okay with my request. I also know that there are times when James, a very solid climber, feels that I go a little overboard on safety. But he understands that during rescues we have to take fewer risks than we might while climbing recreationally. This is partly because if we become victims, we endanger our team as well as the people we're trying to help, and also because professional rescuers must limit their exposure to risk if they want to have a long career. And finally, we can't expose our agency, the Salt Lake County Sheriff's Office, to unnecessary liability.

Descent to the Saddle

As James walks toward our route, he points out footprints in the snow, leading toward the steep slope.

He sets up the first rappel station by tying one end of the rope around a tree. He uses the end of the rope rather than clipping the middle of the rope into a carabiner, so we can use the full two hundred feet. Of course, it means we won't be able to pull it down after us and use it again.

James clips into the rope and rappels down the steep snow-covered slope. When he reaches a moderately sized tree near the end of the rope, he yells, "Off rappel!" I rappel down to him, bringing two more ropes.

As James and I meet at the second rappel station, we see headlamps from the victims in the distance. Moments later, they see our headlamps and begin yelling. Although we can't make out words, we do hear near-panic in those voices.

From this second station, James rappels down the steepening face, following a trail in the snow left by the climbers. I call down to him, asking if the rope reaches the ground. At some point, he realizes that the victims' route is too far to the east and is moving him onto the near-vertical face. Still on very steep snow, he's able to "batman" about fifteen feet up the rope so he can head west toward the saddle.

We will later learn that the trail James was following was indeed that of the victims. They had been coming down the route we just rappelled, albeit in daylight and without ropes, and took off their snowshoes for the steep descent. Two of the men descended first, unintentionally traveling eastward, where they reached a fifty- to seventy-degree rock face covered with snow and ice. One of the men managed to negotiate this steep face. The second man slipped and tumbled down the wall and into the couloir. The first man called up to the five remaining climbers, telling them to traverse to the west and take a less steep route. As often happens across distances in the mountains, they misunderstood his instructions and headed farther *east,* taking an even steeper route. In the process, two more climbers fell, tumbling hundreds of feet into the couloir. The remaining three climbers were able to negotiate the descent safely.

As I'm waiting to move onto the next pitch, I hear Command planning to insert a third team. They have selected two good rescuers, but this terrain is especially challenging. And given the continuing panicked calls for help, we are going to need the most experienced medical people possible. I take out my cell phone, pleased to see that I have coverage, and call Darren. I tell him I think we should insert only one more team, to limit the chance of a rescuer getting hurt, and that I'd like that team to be Andy Peterson and Tom Moyer. Both are very experienced in the alpine environment, and both are part-time patrollers. I realize that my request will leave some people disappointed; I also realize that my job is to do everything I can to make this rescue successful.

After descending the second rope, James calls, "Off rappel!" over the radio. I clip into the rope and start my descent. It immediately drops into a vertical, narrow V-shaped notch. A dusty white trickle of snow spills down on me from above like a waterfall. My hood is on my head and I try to lean forward to avoid having the snow hit my face, but my backpack is trying to pull me over backward. I'm feeling surprisingly happy. What a cool thing to be doing at midnight!

I can see where James's track splits, first following our victims onto the ice-covered wall and then returning to head west. I tie off the rope to my harness so I won't descend farther, and traverse right. There's a small ridge here, formed by a fallen log and a lot of snow, which I struggle to get over. It's as if I were mounting a tall, slippery horse. I can get my leg up on the log, but I can't quite get my weight up and over it. I take my pack off, push it over the snow-covered log, climb over, and put my pack back on.

The soft snow, heavy pack, and cumbersome winter gloves combine to make the simple tasks—tying knots, unbuckling my pack, using my radio—awkward and time-consuming.

James has rigged the third fixed rope from this point, and I rappel down to join him on the saddle between the two peaks.

The wind can really howl between the two mountain peaks, carrying snow from across the valley, and it has created an overhanging cornice on the lee side of the saddle. The cornice has grown to the size of several minivans. Perched on the edge of the saddle, it looks as if it is just waiting to snap off and tumble down the couloir onto the climbing party below.

I use my radio to call the four rescuers above us and tell them about the cornice. I don't want a rescuer rapping down onto it and being the final straw that causes it to break off— especially since James and I will be in the couloir below.

The Ground Crew's Role

Although our packs are stuffed with medical and rope rescue gear, we are going to need more. A team of people back at the command post is scrambling to load duffel bags with additional gear. Rescuer Vickie Ashby, an avid hiker and climber, is acting as the logistician, coordinating what goes into each bag. The bags are then flagged with surveyor's tape so we can identify each one and can learn, via a radio call, which supplies are in each bag.

Shortly after we arrived at the command post and as soon as we realized the scope of the call, a deputy drove to a nearby department store and bought eleven sleeping bags, which are now being added to the duffel bags.

Although there are now four rescuers on the mountain and two more inbound, we're just the tip of the spear. The crews in the valley who are acting as dispatchers, running air operations and logistics, and dealing with the media are the real driving force—and our lifeline.

On every call, it's almost inevitable that there will be people "stuck" at the command post. On this call, there are many qualified rescuers itching to go on the mountain; that they can't leaves them anxious and frustrated. It's understandable—I would be squirming if I were at the command post, sitting on my hands. Leaving a rescuer at the trailhead is a bit like tying Kobe Bryant to a basketball standard and making him watch his team in the playoffs—excruciating. To the rescuers, being left on the asphalt feels like punishment, but the situation is what it is—we have to limit the number of rescuers who are exposed to hazards, and keep rescuers on standby in case the teams on the mountain need help.

Gear Drop

Life Flight is inbound to the saddle where James and I are waiting. They are bringing additional gear, including the eleven sleeping bags, food, a stove, and more medical supplies.

As the helicopter comes in for the drop, I move behind a tree. The bird hovers about thirty feet up, just above the treetops, with the door open. The rotors envelop me in a complete whiteout of blowing snow. Not wearing my goggles, I can't see a thing; all I can do is press my back against the tree and hope the gear doesn't land on me.

The duffel bags drop and sink deep in the powder snow. James and I drag them onto the saddle, where we leave them for the two teams behind us—we're anxious to get to the patients and provide initial medical care.

As James and I are moving the bags, we continue to hear members of the injured climbing party calling out to us. At times, their calls sound desperate and panicky. James and I discuss some of the scene management challenges we are about to face. We will be outnumbered; people may be severely injured, some perhaps dying or dead; we may encounter language barriers. I have concerns that the situation may lead to aggressive behavior on the part of the panicked climbers. We agree that we will assert control of the scene quickly to allay the panic. James will later tell me that he was a little hesitant to be the first one down, because he was unsure what he would find. I can't blame him.

The injured people are still several hundred feet below us, down a steep snowfield. It isn't as steep as the terrain we just descended, but it is still dangerously steep—similar to a double black diamond at a ski area. Unlike the slopes we just descended, if we fall here, we will slide but not die.

A recent storm left a foot of light Utah powder. This morning's report from the Utah Avalanche Center said the avalanche hazard is low today, though it is "moderate on steep upper-elevation wind-affected slopes, mainly on north through southeast aspects." That's exactly what we are on: upper-elevation, wind-affected slopes facing northeast.

James sets up the fourth rappel station at a tree on the saddle, purposefully selecting one that's well away from the cornice. This is a single, two-hundred-foot rope. He throws it down the slope, but the slope isn't steep enough and the rope just plops down into the soft snow.

James clips the rope into his rappel device and begins to work his way down this final pitch. The deep, wind-drifted snow slows his descent, so that he's basically cutting a trench into the steep snow as he rappels.

When he reaches the end of the rope, he calls, "Off rappel!" and I clip in and rap down the hip-deep snow trench. The rappelling is certainly easier for me because James has already cut the first path and pulled the rope down, but I still sink into the snow and have to lean back and push with my feet to make downward progress.

Descending this pitch feels more like being tethered then rappelling, because we can't hang our full weight on the rope. This final pitch has a sort of spacewalking feel as we move silently through the sound-damping powder, illuminated by the moon, bundled in winter gear, wearing our large backpacks and tethered by a thin lifeline.

As Keith is continuing down the third fixed line, Dan is on the second pitch, near the V-slot where I was dusted by the waterfall of snow. Due to Dan's overstuffed backpack, he starts to invert on the rope, until he is hanging nearly upside down. Using his radio, he yells something like "Keith, where are you?" Keith tells him he is down below. Dan calls back, with obvious concern in his voice, "I need help!" The word "help" stings; are we going to have to rescue a rescuer? Being far below him, there isn't much Keith can do, short of ascending the ropes.

A few minutes later, as Life Flight is coming in for a second gear drop, the command post calls to the pilot, who doesn't answer. They call again, and the pilot says something like "I think a climber is in trouble. Do we need to keep the radio clear?" Dan answers something like "Thanks, but I'm okay now." That's a relief for everyone; we already have multiple patients on the mountain.

As we are working our way toward the patients, Tom and Andy, the final team, are flown in. This time the pilot manages to land the helicopter on the summit.

Tom and Andy have the advantage of a summit landing, which shaves hours off their climb to the patients. They have the corresponding disadvantage of being loaded with a huge amount of gear, including two oxygen tanks that we couldn't safely drop from the helicopter. They will be working like mules, because they also have other gear waiting for them at the saddle between the two peaks.

Reaching the Patients

When I reach the end of the rope, James is waiting for me. We've rappelled almost six hundred feet. It's a little before one a.m.—almost four hours since the helicopter dropped us off. We plow through a few hundred feet of snow to reach the climbing party. I can see several people lying on the snow.

I report in: "Command, Team One has arrived."

After the accident, the uninjured climbers shoveled a small shelf in the snow a short distance from the accident site and out of the main avalanche path. They moved their injured friends to the shelf and added clothing. Four people are now huddled together on this shelf.

Although the climbers didn't have many postaccident options, they made good decisions by getting everybody out of the avalanche path and onto the snow shelf, calling for help, and keeping the patients as warm as possible. I'm glad none of them hiked out on their own—I'm sure it was tempting.

There are seven people ranging in age from forty-one to sixty-three. Some in the party are experienced hikers, having summited most of the major peaks in the Wasatch Mountains, but they got in over their heads on this one. I don't think many people, even veteran climbers, would consider this descent without ropes, even in daylight.

Assessments

Since there are several patients and only James and me to provide medical care, we need to triage the patients—sort them to determine the order in which they should get medical care. It may seem harsh, but the triage process doesn't necessarily give priority to the most seriously injured patients—it gives priority to the most "salvageable" patients. Anyone so badly injured as to be near death and unlikely to survive goes to the bottom of the list, because medical responders need to focus on those who can be saved. I haven't found myself in the position where I had to let a patient to die while I treat another patient—let's hope I don't tonight.

James begins the assessments with pencil and paper in hand. He obviously can't take the time to do a full assessment of every patient. Even if he were to spend only a few minutes to do a rapid trauma assessment, get a brief medical history, and get vitals from each person, it would take at least a half hour to assess the seven climbers. Instead, he does triage-style assessments, spending less than a minute with each patient.

He comes back to me with his scrap of paper in his now-numb hands. He briefly tells me what we have: four patients, three with traumatic injuries and at least one with hypothermia.

Like med students on grand rounds, we proceed to each patient, with James telling me their chief complaints and his findings. I ask a few questions of each patient, do a quick inspection, and we move on. To keep track of the patients, James has numbered them one through four, in the order they are arranged on the snow.

The other three members of the party are walking around. We will check their medical conditions later. For now, if they can walk, they're healthy enough that they don't get our immediate attention.

Knowing that the rescuers in the valley will be anxious for a status report, I radio in: "Command, stand by for a medical report." I list each patient by number, age, gender, and key medical information.

Unbeknownst to me, all our radio traffic is being monitored by the media. Tomorrow my brother in California will tell me he heard a portion of my medical report on TV: "Patient number four is a forty-three-year-old female. She has a left arm injury. It appears to be a closed left humerus facture."

Command acknowledges my report. A few minutes later they call and tell me that Fire is asking for blood pressures. I hesitate and then reply that due to "environmental conditions," we don't want to expose the patients any more than necessary and hence, we will not be taking blood pressures. I'm a little concerned that I am refusing to provide the medical care that medical control is requesting, but it's just too cold to justify removing the bulky winter jackets and exposing already compromised patients just to get blood pressure readings. For now, if we can feel a pulse on the wrist, the blood pressure is good enough.

After giving the initial medical report, we get right back to the patients. As I look at our four patients laid out on the ground, I realize how far off I was to worry about aggressive behavior. These patients are a responder's dream: as polite and cooperative as anyone I've ever met. They repeatedly tell us how much they appreciate our coming to help them.

Compounding the injuries is the cold. It's *wicked* cold. Someone in the command post called the National Weather Service and gave them our GPS location and elevation. They were told the temps are between zero and five degrees Fahrenheit.

A rescuer coming down the final rappel keys his mike and says, "Heads up. There's a sleeping bag coming at you." James and I look up and see a sleeping bag tumble down the hill, fly past us, and disappear into the darkness.

I go to the most critically injured patient, a woman. She's complaining of severe pain in her left thigh; pointing to her femur, she explains how much it hurts. She tries to move the leg and winces in pain.

As I take out my trauma shears and pull up a fold of her almost-new Gore-Tex pants, she says, "No, please don't cut my pants." *Snip.*

I then pull up a fold of her long johns. In a pleading voice, she says, "No, not my long underwear, too."

I've learned that if I need to expose an injury, the longer I hesitate, the more adamant the patient will become, insisting that I not cut their (usually expensive) outdoor clothing. If I cut right away, they stop complaining. I do avoid cutting clothes when I can, but in this case it's pretty clear that I need to expose her thigh immediately. *Snip.*

Using my headlight and hands, I check her thigh for "decap beetles." No deformities, contusions, abrasions, punctures/penetrations, burns, lacerations, or swelling. There's some tenderness when I press on her leg, but that may be referred pain from another location. Her leg may be injured, but I don't see signs of an injury.

The extreme leg pain, with only a minor pain response to touch, makes me consider that this may be sciatic nerve pain from a spinal injury. I call the command post and tell them that the patient who I reported as having a possible femur fracture appears instead to have a possible spinal injury. Fire medics working with Command tell me to also suspect a pelvis fracture. In the coming days, I will learn that they nailed it.

Considering the likelihood of a spinal or pelvic injury, Tom, who recently arrived, suggests having Life Flight drop the full-body vacuum splint at the saddle. He then climbs five hundred feet up the steep snowfield, gets the large duffel bag containing the beanbag, and rappels down.

Spine and pelvis injuries can be life threatening. Here we are on the north side of a mountain, thousands of feet above the valley, in January, on six feet of snow, in single-digit temperatures. How in the world are we going to get this woman to definitive medical care?

I call the command post: "There is still a possibility we may want to bring in a litter tonight and evacuate one of these patients." However, I know that having a toboggan delivered from a helicopter isn't easy. We can't drop it from the helicopter, and we can't use the hoist at night. We would need to have the two-piece toboggan placed inside a helicopter and flown to the summit. Climbers would then need to rappel six hundred feet with the litter to reach the patient. Not only would it take hours to get the toboggan here, it would take several more hours to drag her out using skis and snowshoes, which would also need to be flown in.

We quickly rule this option out, at least for the time being. It is after one a.m., and the sun will rise in less than seven hours. Life Flight has agreed to have a fresh flight crew on scene at daylight to hoist the patients.

Although the weather forecast says clear, calm, and cold, Andy and I discuss our options should we be unable to use the hoist in the morning. What if the weather changes? What if the hoist-equipped helicopter has a mechanical problem or is dispatched to another call? We talk through the process of evacuating all four patients using toboggans. If we borrow a second two-piece toboggan from Brighton, we can have a team of rescuers dragging a patient out while another team rappels in with a toboggan. It's doable, but we agree that it will take twenty-four hours or more to transport all four patients and will require many more rescuers.

While we are doing assessments and treating patients, other rescuers use shovels to enlarge the platform in the snow. We then lay out the beanbag. The six rescuers lift the woman in a bridge lift, keeping her spine as stable as possible, and put her on the bag. Her voice makes it clear she is in terrible pain. We pump out the air to make the splint rigid. We then put the bag on a sleeping bag and cover it with two more.

The next patient up for care on James's triage list is a sixty-three-year-old man. He looks much younger than his age and is in good physical condition.

He reports pain below his left collarbone, has difficulty moving his left arm, and tells us it hurts to take a breath. I unzip his jacket and reach my gloved hand under his clothing. He doesn't appear to have a fractured clavicle, but he has obvious pain just below the collarbone.

He is stoic and only answers direct questions. His answers are quiet and almost a mumble. James and I discuss his lethargic behavior and the possibility of a head injury. When we try to get more specific with our questions, he tells us we should "take care of the women." I can't tell whether his terse, solemn responses are the result of a language barrier, a head injury, or just his personality. He seems despondent and resigned to the seriousness of the situation.

The location of the pain—upper chest—and the fact that it hurts when he takes a deep breath lead us to consider a rib injury. And it's likely that the chest trauma also resulted in a lung injury. Although I don't want to expose his chest, I need to listen to his lungs. I take out my stethoscope and put the end under his shirt. I listen first to his "good" right side, then move to the injured side. I do this repeatedly at different locations on his chest and back. Each time I ask him to take a deep breath, he winces slightly. He repeats his mumbled statement: "I'm okay. Help women." His lungs sound clear, with bilateral breath sounds, and I don't hear the crepitus of crunching bones, but I suspect a significant chest injury and a possible head injury.

Lung injuries can leak air or blood into the chest cavity, where it then occupies the space that would normally be used by the lung. Eventually, the space for the lung can be reduced to the point that the lung can no longer fill with air. The treatment is to penetrate the chest cavity with a needle and allow the air, or blood, to be released. The fact that I don't have the training or equipment to deal with a serious lung injury crosses my mind.

Looking at this patient, I'm convinced he is one tough nut, who doesn't complain. That said, I worry whether he will make it through the night. Lung and head injuries can quickly deteriorate, progressing rapidly to death if not treated. And treatment for his injuries must happen in a hospital. The best we can do here is monitor him closely to see if he begins to deteriorate. If he does, we will provide as much care as we can, such as maintaining his airway and providing oxygen.

The third patient appears to have a severe fracture of her wrist and possibly her humerus. I can see blood caked on the sleeve of her down jacket.

I try to expose the injury, but the opening on the sleeve of her coat is too tight. The bleeding, pinpoint pain, and the patient's guarding of the injury are typical signs of an open fracture. And open fractures should be exposed. I take out my scissors to cut her sleeve, then decide that in this case, the insulation of her down coat is more important that my seeing the injury. The bleeding has stopped and is not a life threat; we will splint the arm and keep her warm.

The fourth and final patient says she is uninjured but cold. She appears to be suffering from hypothermia and possibly frostbitten toes.

We lay out sleeping bags and prepare to lift her without moving her spine. When we do, her hiking boots remain glued to the snow. I grasp both feet and give a strong pull, expecting to dislodge them. I should know better; they don't budge. Even though they are only an inch or two into the snow, they are frozen like toothpicks in an ice cube. A rescuer digs out her feet, and we move her onto a sleeping bag.

Normally we don't feed patients with traumatic injuries, because our transportation times are usually short. But these patients haven't eaten for more than ten hours. We give them energy bars and water heated on a camping stove.

National News

About every half hour, we hear the sheriff's office dispatcher call the command post on the radio to tell them that yet another television station wants to speak with somebody about the rescue. They have been getting calls from all the local TV stations as well as CNN, MSNBC, FOX in New York, and others. It's a little weird to be on a mountain and hearing that somebody in New York is calling to find out what we're doing. Inside the command post, our fellow rescuers have been watching the story unfold on TV. In this high-tech world, rescuers can *be* the news and *watch* the news at the same time.

Everybody's Gotta Go Sometime

The woman with the possible spine or pelvis injury bashfully tells Andy that she needs to pee. She is embarrassed, quiet, and polite.

Normally, prehospital care workers never remove spinal immobilization, but we have to consider the tradeoffs between spinal immobilization and getting a patient wet in freezing temperatures. Andy pulls me aside and tells me he has real concerns that getting her wet in this environment may bring on life-threatening hypothermia.

We decide to undo the beanbag. Four rescuers lift her up while maintaining spinal stability. Two other rescuers slip a thin, Mylar space blanket under her, and we put her back down. We then form the space blanket so the urine will run down the blanket and onto the snow. The other rescuers retreat to give some semblance of privacy. Andy, courteous and discreet, remains to ensure that we don't worsen her medical conditions. She goes, and we repackage her in the beanbag and again cover her with sleeping bags.

Six Leave

After stabilizing the patients, we decide to have three rescuers—Tom, Dan, and James—hike out with the three uninjured climbers. Andy, Keith, and I will stay with the four patients. It's three a.m.

When the climbing party tumbled down the ice-covered wall, their equipment was strewn across the snow slope in a giant "yard sale." Although the rescuers brought in a lot of gear, we didn't bring in skis or snowshoes to travel on top of snow; we were equipped strictly for winter climbing.

The rescuers who are leaving climb to the base of the wall to scavenge extra pairs of snowshoes.

As they are leaving, Tom offers me his puffy down coat. Although I don't want to rob him of his jacket, Tom assures me he'll be plenty warm on the hike out, and I'll need it to spend the night up here, high on the mountain.

The six people in this group reach the road four hours later, at seven a.m. They are met by three ambulances, where they are checked and released. We get a radio call telling us they made it out safely and the climbers are okay. Three down and four to go.

Winter Bivy

The patients are resting on top of three sleeping bags. Much of the need for insulation is from the snow, although tonight the snow is warmer than the air. We have five additional bags spread on top of them

Andy checks the patients' vitals every twenty to thirty minutes. He wakes them up with a cheery "Good morning! This is your three thirty a.m. wakeup call! How are you doing?" He asks if he can get them anything—not that we have much to offer. They are very polite, especially considering how Andy keeps interrupting their sleep. He asks each patient if they know where they are and if they can tell him what happened. He rechecks their pulse and respirations.

Andy's humor is entertaining, but he has an ulterior motive. A relaxed demeanor when dealing with a seriously injured patient helps the patient stay calm, allowing the patient's heart and lungs to do more important things than indulge anxieties. Of course, the real point of his wakeup calls isn't to see if he can get them anything; the point is to recheck their mental status and vitals to see if anything is changing.

Our medical supplies are limited. We have enough supplies to start one IV and to give that person up to two liters of fluid, but the fluid is so cold, it might drop their body temperature. And keeping the fluid from freezing in the tubes will be a real challenge. We put the IV supplies under the sleeping bags to keep them warm, but I hope we don't need them.

We have two small bottles of oxygen—good for forty minutes of high flow. Andy and I agree that even though we would normally administer oxygen to the woman with the spine/pelvis injury and the man with the chest/head injuries, our supplies are so limited that we don't want to use up precious oxygen yet. If we notice a change in mental status or vitals during the frequent checkups, that patient gets the oxygen, which may slow the deterioration. We will then triage the order in which we transport the patients.

Keeping Warm

Andy walks back after one of his medical checkups and tells me, "We've got them so warm they're almost sweating!" That's good, since we don't need hypothermia or frostbite complicating their existing injuries.

It's not as easy for Keith, Andy, and me to stay warm. Keith is wrapped up in a skimpy sleeping bag, and Andy and I try to share the remaining bag. At some point, I start digging a hole—mostly to warm myself with the exertion. Then Andy gets up and starts making a road in the snow, leading from the patients. I join the public works project, shoveling until my back aches, and then lie down on the sleeping bag. When I get too cold, I go back to shoveling.

Our road grows and we add a small cul-de-sac at the end. Andy starts building a curb and gutter while I carve faux stone blocks on the wall alongside our road.

Along with physical exercise, enthusiasm and purpose are helpful when trying to stay warm. If I were skiing or hiking in five-degree weather and staying up all night, I would be frozen. But the adrenaline and challenge, the need to stay focused on our patients, and a little shoveling help keep me relatively warm.

The next morning one of the TV reporters will say, in a rather apt slip of the tongue, "Rescuers cuddled—I mean *shoveled*—snow all night to stay warm." When Andy and I were wrapped in the single sleeping bag, I suppose that could qualify as cuddling, but with such scant covering, we were a lot colder than when shoveling.

In one of the duffel bags we find a military MRE—a "meal ready to eat." That sounds good—we're hungry and ready to eat anything that isn't an energy bar. But the long instructions are in tiny print, and the sad truth is, we can't figure out how to cook our dinner. It's easy enough for a soldier in combat, but not for us, so we put the MRE aside.

We have been using a small camping stove to heat water, but eventually we're too tired to bother. Andy tells me he's out of water, and asks if I have any. I've been keeping a one-liter Nalgene bottle under my—well, Tom's—down parka to keep it from freezing. I hand it to Andy, who takes a sip, looks at me, and bursts out laughing—the water has frozen into slush.

Every hour during the night, a deputy in the command post calls us on the radio: "Team One, Command. Are you code four?"—jargon for "Are you okay?" We reply, "Yes, we're code four. No change in our patients' condition. Thanks for checking." Though it may sound a little corny, it's nice to know we aren't "alone" on the mountain.

Getting Out

The sun rises a little before eight. Between shoveling to stay warm and the periodic assessments, we never got to sleep.

As the stars fade into the gray dawn, I can see the pile of gear we brought in last night. I'm glad the pilot was able to insert us and all the gear. Without the helicopter, we would have spent at least four more hours reaching the patients, and we would have had minimal equipment—certainly not eleven sleeping bags. Without the helicopter, I don't know how many of our patients would have survived the night.

Another page goes out for more rescuers, to help with the inbound patients and rescuers. There will also be a lot of gear to clean up and store—one of the less glamorous, yet vitally important, tasks that fall to emergency workers.

```
S&R Callout - Overdue hikers Mt.
Olympus. Meet 600 at the Command Post
5600 S Wasatch Blvd. Cs 6223 (eom)
7:59 01/22/06
```

Andy, Keith, and I have a brief discussion to triage our four patients. We will fly the woman with the spine/pelvis injury out first, then the man with the head and chest injuries. Next is the woman with the fractured arm, and last, the hypothermic patient with frostbitten toes.

In preparation for the hoisting and the pending manmade whiteout, we stuff our personal gear in our backpacks and put the team gear back in the duffels. Then we secure the backpacks and duffels behind a small tree.

We can see the helicopter flying toward us from the east, with paramedic Mike Quinones dangling below. Hanging below Mike is a large red duffel, containing a second beanbag. This will allow us to package the next patient while the first one is being flown out. Behind Mike are two tiny specs—news helicopters hovering in the distance.

As Mike nears the road that Andy and I built during the night, he extends his arm in the direction he wants to go. The nurse in the helicopter, Judi Carpenter, is leaning out the open door and relaying Mike's hand signals to pilot Craig Cowley while she lowers him onto our little road and cul-de-sac. Craig is the brother of Brent Cowley, the pilot who died on the other side of this same mountain while helping our dehydrated patient.

The three of us lie over our four patients, shielding their faces from the windstorm and keeping the sleeping bags from blowing away in the prop wash.

Mike unhooks from the cable and holds the hook to his side in his outstretched arm to show that he's clear and that Judi should raise the cable. Then the four of us move the first patient, who is already in a beanbag, into the hammocklike sack. As we are carrying her to the cul-de-sac, I reflect on what her life has been like for the past twenty-four hours. What began as a fun and challenging mountaineering adventure turned into a traumatic event that stranded her in the mountains for fourteen hours, the last eight of which she has spent immobilized in a vacuum splint.

Mike calls Craig on his radio and says we're ready to hoist, and the helicopter flies in. Judi again guides the pilot to position the hook in Mike's hand, and he connects the hook to the straps on the vacuum splint and to his harness. And with a wave of his arm, he and the patient are rising into the sky.

The helicopter flies around the mountain to the ambulances and a second medical helicopter, which are waiting at the command post.

While this is happening, we package the next patient in the beanbag, and when the bird shows up, we repeat the process. We do this until all four patients are out.

After all the patients have been hoisted, the three of us discuss whether we'd rather hike out or be hoisted. It's a no-brainer. But we also know that being flown out is a bonus and not guaranteed. Life Flight did a stellar job of inserting us last night, dropping off additional gear, and getting the patients safely back to civilization. We couldn't ask for more, and there will be no hard feelings on our part if they don't fly us out.

I call Command and ask if they want us to hike or be hoisted. I let them know we're okay either way but that if we're hiking, we need skis or snowshoes dropped at the top of the couloir, where we will climb to retrieve them. After a short delay, Command tells us that Life Flight will be glad to hoist us, but they don't have enough fuel to get us all without making an intermediate fuel run.

Andy, Keith, and I collect all the gear, including our backpacks and the remaining equipment from the climbers, and put it in a big pile. It's about four feet square and weighs three hundred pounds.

The helicopter flies in again, drops Mike off, and leaves. Andy puts on the FAA-approved harness—government regs don't allow us to fly with our trusty climbing harnesses—and Mike connects Andy to himself with a short piece of rope. The helicopter returns with its hook dangling, Mike clips them on, and up they go. After dropping Andy off, the chopper comes back for Keith.

I am now alone on the mountain. Our little commune no longer exists, and our road and cul-de-sac, and Andy's curb and gutter, have been destroyed by the repeated hoistings.

Although we didn't have to perform Herculean medical treatments, I feel a tremendous surge of satisfaction knowing that all our patients survived the night and have reached definitive medical care. And as always, there's the lift that comes with handing off the responsibility.

I watch the helicopter coming for me, with Mike dangling below. I'm ready to get out of here. I put on the harness, and up we go, to hang just outside the door.

The adrenaline buzz of dangling a thousand feet above the crags, from a cable half the diameter of a pencil adds to the enjoyment. I look up at Judi in the doorway, running the hoist, and give her a big smile. She smiles back but quickly restores her game face. After all, she is now the rescuer, focused on delivering another patient to the road.

While hanging in the sky, I realize that I'm so cold I can barely feel my fingers. I let my arms hang loose at my side, hoping to encourage their circulation. I expect that my rescuer, now on his seventh hoist of the morning, must be freezing.

The helicopter swings around the mountain in a large banking turn. Jagged cliffs pass below us. As we round the mountain, the Salt Lake Valley comes into view. Below me, a million people are going about their daily activities, and here I am, hanging in the sky above them. What a great place to be!

Back in Civ

As we approach the command post, I see just how big an operation this has been. While I was bivied on the mountain with two other rescuers and four patients, I was aware of the support we were receiving, but I hadn't quite comprehended the magnitude. There are three ambulances and another helicopter on the ground, the team's command post and a huge incident command RV belonging to the Unified Fire Authority, red paramedic Suburbans with flashing lights, a half dozen sheriff's SUVs, two of which are stopping traffic for my arrival, five or six media trucks with satellite dishes perched atop telescoping poles, and a dozen rescuers' vehicles.

With the helicopter hovering fifty feet above this assemblage of vehicles and flashing lights, Judi activates the hoist, and the paramedic and I slowly descend to the road below. I feel as if I were being lowered on a thread into a sea of activity: lights and media, firefighters in turnout gear, cops in uniform, and people being interviewed by TV reporters. This is a different world from where I spent the night, on the back side of a moonlit mountain. Somebody unhooks me from the cable, which then rises into the sky. I look around and see dozens of people looking at me.

Sgt. Thad Moore comes up, shakes my hand, and congratulates me. Then, sounding a little apologetic, he tells me he's having everybody checked out medically, and walks me toward a waiting ambulance as I flash my teammates the victory sign.

Steve returns to civilization

After being checked out, I walk back to the command post. I'm stopped by a cute reporter and give a quick TV interview.

I sit down on a bench seat in the command post, and somebody microwaves a frozen Hot Pocket for me. After all that healthy work, you'd think we'd have healthier food. Oh, well, at this point I would be happy with gruel.

I look at my teammates, including those who have been stuck in the command post. After a mission like this, there is a tremendous feeling of bonding with my fellow rescuers that's a little hard to express, to each other or anyone else.

I look up at the TV mounted in the front of the command post and watch clips of the patients being hoisted into the sky. It is a surreal time-warp.

There isn't a lot of excitement inside the command post. Everyone is glad the mission was successful. My teammates, stuck in this aluminum box, look bored and left out. I am whipped.

Then the shivering begins. Just a little at first, and then it sort of takes over. Laurie Jess brings over several blankets and puts them on me. I feel confused, though I don't say anything. Why is Laurie putting blankets on me? Why am I shaking? Why can't I stop? I don't *feel* cold. In hindsight, I was suffering from hypothermia. Then the warm command post and the calories in the Hot Pocket gave my body enough energy to start my muscles shivering. It's a good thing Tom left me his down jacket last night.

Mission Completed

The final page comes at two in the afternoon, twenty hours after the initial call:

```
Cancel S&R. Mt Olympus rescue is
completed. Per car 600 thanks for
your help. (eom)
14:01 01/22/06
```

Darren Westerfield, Mike Brehm, and Ric Bruce were flown in to retrieve our gear. They landed on the summit, which Mike, seeing it in daylight, described as "challenging" and "the size of a small dining room." The LZ was so small, in fact, that when the helicopter returned with his teammates he was glad he was securely tied off, so the rotor wash wouldn't blow him off the peak. Sometimes things look better in the dark.

Untying our fixed ropes from the trees, they tied them to new ropes they had brought, looping the doubled rope around the tree anchor so they could then pull it down after rappelling. When they got to our bivy site, they put the three hundred pounds of duffel bags and backpacks in a cargo net, and the DPS helicopter long-lined it all on a fixed rope down to the valley. As Darren, Mike, and Ric snowshoed to civilization, all that remained were indentations in the snow.

Thanks, or Not

This rescue got an inordinate amount of media attention. The attention is okay, but for most rescuers its real benefit is to give the people we know, our friends and family, a little better understanding of what we do.

When I got home from the call, still in a daze from the event, I received calls from both the *Today Show* and *Good Morning America*. Both shows scheduled interviews for the next morning, but they called me later that night to cancel because the victims didn't want to appear on TV.

Media madness
(Visit MountainResponder.com/Photos and enter Story Code MOK.)

A few months later, the seven members of the Korean Alpine Club came to one of our monthly Search and Rescue meetings. They brought a huge cake and a plaque thanking us. Dressed in suits, they gave a speech thanking us and posed for pictures. In my informal T-shirt and shorts, I looked a little out of place. Their gratitude was nice; it also made the rescuers a little uncomfortable.

It's odd what makes different people feel out of their element. Most people would be uneasy about climbing at night on a snow-and-ice-covered mountain or dangling under a helicopter. Rescuers, on the other hand, are uncomfortable being thanked.

The club also hosted a Korean barbecue for the members of the team a few months later. Unfortunately, I was out of town and couldn't attend. It was clear that these folks truly appreciated our help.

Sometimes the appreciation is less obvious. I remember spending most of a night carrying a litter bearing a teenage girl with a suspected spinal injury. When we reached the parking lot at the trailhead in the early morning, fifteen to twenty members of her family, an ambulance, and several paramedics were all waiting for her to arrive. They swooped in, loaded the girl into an ambulance, and took off, and within minutes it was quiet again as the rescuers silently packed up their gear. I guess it makes sense that nobody said thank you. They were focused on an injured person they loved. In the same circumstances, I might react the same.

Sometimes the appreciation is ours—for our friends and families, whose lives are repeatedly disrupted by the misfortunes of others as our pagers go off amid dinners, birthdays, and romantic moments. A few years ago, three rescuers and their significant others were settling into soft theater seats to watch movies from the Banff Film Festival—films of epic outdoor adventure and on-the-edge risk taking. When I felt my silenced pager vibrating on my belt and saw my teammates squirm, I knew that calamity was calling. As we left our seats and headed toward the avalanche-prone mountains, my girlfriend had to chase me up the aisle and remind me that she would need my car keys to get home—my mind had already shifted gears into rescue mode.

But despite the sacrifices—both ours and our loved ones'—and whether or not anyone remembers to bestow a word of thanks amid all the emotional intensity, I know how privileged I am to have participated in this and hundreds of other rescues. During that time, I have seen well-honed skills and teamwork conquer tremendously challenging situations, and I'm proud to have been a member of the team.

Epilogue

I was on the Salt Lake County Sheriff's Search and Rescue Team for six years when a new Sheriff, Jim Winder, was elected. He had been involved with the Search and Rescue Team as a deputy about ten years earlier, and I was looking forward to his help as we continued to strengthen the team. Unfortunately, that would not be the case.

Shortly after taking office, he replaced the lieutenant and two sergeants who had supervised the team. Following Sheriff Winder's instructions, the new guys proposed a sweeping reorganization that would replace the team's leadership structure with one led by sheriff's office employees. The plan proposed that rescues would be done by ten-person squads, each of which would only include three or four volunteer members. The remaining six or seven rescuers would be deputies and firefighters with minimal or no backcountry experience. The team would no longer work and train together as a unit but would instead work in separate squads.

The team members were baffled by the scope of the changes and by the complete lack of involvement by anyone with mountain rescue experience. The two sergeants newly in charge had spent nine months creating the plan *without ever participating in a single rescue or training* and without consulting team members or the team's elected board. In a team vote, only two of the team's thirty-five members said they would remain on the team under the proposed plan.

We were proud to be a part of the sheriff's office, and we never questioned who should ultimately make the decisions, but we seriously questioned the wisdom of a plan that relied on inexperienced rescuers and leaders. The plan was made by people who had no clue what my teammates and I did or why we did it. Sheriff Winder and his men seemed to believe that anyone, given a few months of training, could learn the necessary mountain rescue skills—never mind that many of us who had been at it for years or even decades felt that we had plenty to learn.

Over the following months, a committee of team members worked in good faith with the sheriff's office to resolve our major concerns. During this process, we proposed using mediation to find a resolution. Sheriff Winder's response was that "the Sheriff's Office is a paramilitary organization" and that there was "no need for mediation."

As much as I loved rescuing people in the mountains, I had a difficult decision to make. As commander of the team, my core objectives were to improve safety, complete missions successfully, and strengthen the team. Unfortunately, I felt that the proposed plan would make us less safe, less able to complete missions successfully, and would weaken the team.

The "final, non-negotiable" plan was presented on March 3, 2008. Within a week, eleven rescuers, including me, resigned. With these resignations, the team lost more than two-thirds of its active team member experience.

Our resignations ultimately resulted in the sheriff reassigning the two sergeants who created the plan. Adjustments were made to the plan, but it continued to be designed and implemented by people who didn't seem to understand what we rescuers did, why we did it, or why so many of its most experienced members resigned. They seemed to be more interested in power and authority than in the well-being of people who get lost or injured in the mountains. The sheriff's staff was replacing the democratic, transparent culture of the team with an alien one that valued obedience over competence, and which many of us just could not support.

The volunteer team had executed rescues for fifty years without mishap or serious conflict with the sheriff's office. We were a self-healing, self-training team of dedicated rescuers who were glad to work without pay.

There is obvious irony in the fact that it wasn't rockfall, a helicopter crash, an avalanche, or old age that ended the careers of rescuers with a combined 122 years of experience encompassing 4,124 rescues. Rather, it was a reorganization, designed by people who knew little about mountain rescue. Among those rescuers who resigned were many who appear in this book, including Alan Erdahl, Amy Fisher, Andy Peterson, Casey Sullivan, James Taylor, Laurie Jess, Mike Brehm, Tom Moyer, and Vickie Ashby.

To say that our resignations left me disappointed and disheartened is an understatement. But I remain convinced that it was the right decision—that speaking truth to power was necessary. I remain proud of the rescues that my teammates and I participated in, and of the role that volunteer rescuers continue to play throughout the country.

About the Author

Steve Achelis is the former Commander of the Salt Lake County Sheriff's Search and Rescue Team. During his time on the team, he responded to hundreds of rescues—calls that included more than forty fatalities.

Salt Lake County is unusual in that a million people live within a half-hour drive of some of the country's best rock climbing and backcountry skiing. This combination provided Steve and his teammates with many opportunities for rescues.

Steve Achelis

Steve is an EMT-I, WEMT, Outdoor Emergency Care instructor for the National Ski Patrol, and Wilderness First Responder instructor for the Wilderness Medical Institute. He is also a member of Wasatch Backcountry Rescue and the Brighton Ski Patrol, where he patrols on telemark skis or a snowboard.

When he isn't playing in the mountains, Steve is a software entrepreneur. His current technology-based rescue projects include eMedic.com, RescueRigger.com, BeaconReviews.com, and SmartMedic.com. Steve's current projects are summarized at Iterum.com.

LaVergne, TN USA
19 May 2010
183246LV00005B/22/P

9 781608 441075